Praise for *The Trainable Cat*

"Interesting premise. . . . The goal here is not to get your cat ready for the Big Apple Circus, but to make it easy for you to get your cat to do all the things many cats resist: swallow a pill, go to the vet, take a bath, or stop trying to disembowel your new cat. . . . My favorite tip is this: To get a cat used to a new home, you take some white cotton gloves and basically feel up your cat, then rub the gloves all over the furnishing. Smelling her own scent, the cat will believe she's been there before and calm down. This is genius."
—Judith Newman, *New York Times*

"Do you want your cat to come when you call it, stop destroying the furniture or killing birds, and enjoy taking a walk on a leash? Then this is the book for you. By integrating established principles of animal learning with the latest in feline science, *The Trainable Cat* will enrich the lives of cats and their owners."
—Hal Herzog, author of *Some We Love, Some We Hate, Some We Eat: Why It's So Hard to Think Straight About Animals*

"You can train a cat to do just about anything a dog can do, except a cat may do it better! John Bradshaw and Sarah Ellis illustrate how cats are trainable, but, more importantly, the authors bust long-held myths about cats and cat behavior along the way. As a result, both experienced stronger bonds with their purring pals."
—Steve Dale, certified animal behavior consultant and author of *The Good Cat!*

"*The Trainable Cat* breaks down the myth that cats cannot be trained. Not only can they be trained but training improves cats' quality of life. This book should be required reading for all cat lovers, including all veterinary professionals who work with cats. *The Trainable Cat* will change how you think about cats, improving relationships and preventing behavior problems."
—Ilona Rodan, DVM, board-certified feline specialist and coauthor of *Feline Behavioral Health and Welfare*

"Typically, your cat doesn't care what you read. This is not one of those times. Cats around the world would be ecstatic to learn that you are not one of those people saying 'cats can't be trained.' John Bradshaw and Sarah Ellis are experts in feline behavior, and together they bring clarity to a vastly misunderstood topic. Cat training can help cats lead happier lives alongside the people who love them. And it's fun! Read this book. Your cat will thank you."

—Julie Hecht, MSc, author of *Scientific American*'s Dog Spies blog

"John and Sarah have demystified cat training, making it accessible to all cat lovers—from professionals to owners alike. Clear instructions for training along with real-life anecdotes expose the amazing potential of our domestic cats. This is a must-read for all cat lovers who are interested in providing the best life possible while building a deeply trusting partnership with their feline companions. This book has the potential to save lives!"

—Miranda K. Workman, clinical assistant professor of animal behavior, ecology, and conservation, Canisius College

"I love this book! We often greatly underestimate the capabilities of our pet cats, and *The Trainable Cat* is a thorough yet completely accessible resounding YES in response to the question: Can you train a cat? Training can not only solve behavior problems, but it strengthens the bond between a cat and their humans. This book will enhance the welfare of cats everywhere, and change how we think about what cats can do."

—Mikel Delgado, certified cat behavior consultant, scientist, and blogger

"Fantastic! *The Trainable Cat* is an accessible resource for cat parents and professionals alike. The authors debunk common myths and misconceptions about cats and their trainability, all while equipping trainers with the knowledge they'll need to overcome challenges relating to rewards, motivation, and individual learning styles. I will definitely recommend this book to my clients!"

—Ingrid Johnson, certified cat behavior consultant, Fundamentally Feline

"Finally, a comprehensive look at how operant training can be used to benefit our feline friends by preventing or modifying behaviors that we humans find upsetting. Bonus, it's really fun too!"

> —Jacqueline Munera, certified cat behavior consultant and cocreator of "What Is My Cat Saying? Feline Communication 101"

"I was delighted to read this book. *The Trainable Cat* is intelligent and it's accessible, from discussing a cat's motivations to concepts of learning to training them (not the owner!) using their own instincts, not for tricks, but for grooming, trips to the veterinarian, getting along with the dog or the other cats, taking medication, and so much more. There are many books on cat behavior, many which cover the same material. Not this one. Bradshaw and Ellis's book is original in concept and research, and the authors demonstrate an enormous understanding of a cat's personality—and respect for a cat's needs. There's wisdom and compassion here. *The Trainable Cat* is an important book for both feline specialist and owner."

> —Jane Ehrlich, associate certified feline behavior consultant and owner, Cattitude Feline Behavior

The Trainable Cat

ALSO BY JOHN BRADSHAW:

Cat Sense

Dog Sense

BY JOHN BRADSHAW AND SARAH ELLIS:

The Trainable Cat

The Trainable Cat

A PRACTICAL GUIDE
TO MAKING LIFE HAPPIER
FOR YOU AND YOUR CAT

JOHN BRADSHAW
AND SARAH ELLIS

BASIC BOOKS
New York

Basic Books
Hachette Book Group
1290 Avenue of the Americas, New York, NY 10104
www.basicbooks.com

Printed in the United States of America

First Paperback Edition: October 2017

Published by Basic Books, an imprint of Perseus Books, LLC, a subsidiary of Hachette Book Group, Inc.

The Hachette Speakers Bureau provides a wide range of authors for speaking events. To find out more, go to www.hachettespeakersbureau.com or call (866) 376-6591.

The publisher is not responsible for websites (or their content) that are not owned by the publisher.

Print book interior design by Trish Wilkinson.

Library of Congress Cataloging-in-Publication Data

Names: Bradshaw, John, 1950–, author. | Ellis, Sarah, 1980–, author.
Title: The trainable cat : a practical guide to making life happier for you
    and your cat / John Bradshaw and Sarah Ellis.
Description: New York : Basic Books, 2016. | Includes bibliographical
    references and index.
Identifiers: LCCN 2016019146 | ISBN 9780465050901 (hardcover)
Subjects: LCSH: Cats—Training. | Cats—Behavior.
Classification: LCC SF446.6 .B73 2016 | DDC 636.8/0835—dc23
LC record available at https://lccn.loc.gov/2016019146

ISBN 978-0-465-09371-7 (paperback); ISBN 978-0-465-09649-7 (ebook)

LSC-C

10 9 8 7 6 5 4 3 2

Dedication: to Sarah's beloved cat Herbie

During the final weeks of writing this book, my beloved Herbie unexpectedly died. I would like to dedicate this book to him, without whose inspiration I would not have had the knowledge, skill or power to complete this book. My wish is he leaves a legacy behind—helping to teach many more owners how they can help their cats cope with, and even enjoy, the trials and tribulations that living alongside us brings.

Herbie . . . your paw prints forever walk my heart.

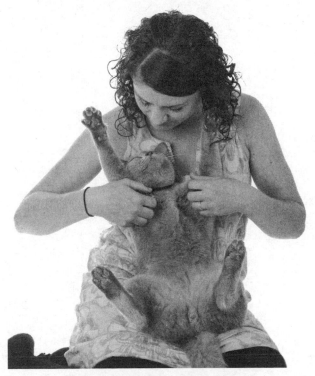

Herbie was a special cat in many ways,
here enjoying Sarah tickling his armpits.

Contents

Sarah's Preface

TRAINING CATS IS SOMETHING THAT I STUMBLED UPON. Now, when I think back to when it first started, I realize I was actually about seven years old—although obviously I did not know at that time what impact this would have on my later life, or that of the cats I would own. My mother bought our family a Burmese kitten. This kitten became the subject of all my affections—forming the focus of elaborate projects for Brownie animal lovers' badges, paintings and illustrations for art competitions and much talked about at school to friends, to name just a few. Claude, as he was known, was very affectionate, food motivated and active—the perfect combination of traits for training. Before long, Claude was being lured over elaborate furniture obstacles with tasty morsels snuck from the fridge and excitedly chasing a wand toy that I dragged at high speed over garden jumps made from my mother's clothesline poles. I think my proudest party trick was to double tap my shoulder, whereupon Claude would jump from the back of the sofa up onto my shoulder and balance there as I moved carefully around the living room to the windowsill. With a double tap on it, he would leap down and rub his face against mine. There is no doubt that the love I felt for Claude was reciprocated—he slept in my bed and in my doll's cots and often accompanied me on dog walks. I lost Claude on my twenty-sixth birthday at the grand age of nineteen—what an imprint he had made on me, as by this time I was already halfway through studying for a PhD in feline behavior.

My research, alongside my professional work with feline problem behaviors, gave me great insight into current welfare concerns associated with modern cat ownership—I discovered just how few cats enjoy going to the vet, lie back and relax while traveling in the car, enthusiastically open their mouths and swallow their worming pill, or embrace the new addition to the family, whether feline, human or canine. As a dedicated cat owner, I felt it only right to try to ensure that my cats, from an early age, had the life skills to be unfazed by such events. I am not a professional animal trainer, but I have been very fortunate to have worked alongside some wonderful trainers who have shared their knowledge and practical skills with me and on occasion, let me assist with training (puppies).

Combining this with my knowledge of cats and learning theory, I began incorporating training into the daily lives of all my cats. Training has helped my cats cope well with the many challenges life has thrown at them. Positive feedback came early on: owners sitting next to me in the veterinary waiting room would remark with surprise at how calm and relaxed my cats were in their carriers, and the vet would comment, "I wish all cats were as good as yours." From that moment on, I decided that I would train every cat I owned as a matter of course.

Woody, the first cat I owned in my adult life, moved house with me many times, even crossing the Irish Sea without batting an eyelid, thanks to the preparatory training I had carried out. Later, Cosmos, another of my cats and one who appears throughout this book, warmly shared his home with several foster cats and subsequently accepted Herbie, who arrived as a permanent resident as a mischievous and playful kitten. A few years later, Herbie and Cosmos were taught to share their home with a canine addition—Squidge the Jack Russell. Knowing I might get a dog at some point in their lifetimes, I had begun training them to accept visiting dogs from an early age, well before we brought home Squidge.

The most recent addition has been baby Reuben, perhaps the addition I was most apprehensive about—after all, there was no way he could be rehomed if it didn't work out! However, I am delighted that

Reuben, now a toddler, is showing promise as the next-generation cat lover in our family and that all the preparatory and ongoing training with the cats has led them to relish his presence; Cosmos will often chirrup at Reuben when he comes in from outdoors and greets him with a rub of his face.

Because training had produced such happy consequences for my own cats, I felt compelled to spread the word—part of this was a year-long series of how-to training articles for a UK national cat magazine. The articles were warmly received and I realized that they had really touched only the tip of the iceberg. I was very keen to do more, mulling the idea of a book on the topic over in my head. It was at this time that I began working with the BBC on their Horizon TV program *The Secret Life of the Cat*, where one of my first roles was to teach owners how to train their cats to wear GPS tracking devices. John, who worked alongside me on the program, witnessed this and as we began to talk about training cats and to share ideas, the concept for a jointly written book was born. While training cats may have all begun unintentionally for me, I've seen such positive impact on cat welfare that I'm inspired to spread this how-to training across the globe.

John's Preface

As Sarah says, the concept for this book came together in 2013, when we met up at "Cat HQ" in the picturesque village of Shamley Green, the location for the BBC's Horizon TV documentaries about cats. I have to confess that until Sarah suggested the idea to me, I had never thought much about training cats. I knew a few people who had trained their cats to do tricks, including one who delighted in his cat's ability to hop up onto the seat of a conventional toilet and use it in place of a litter box (no, we will not be describing how this might be done in this book!). I'd come across "performing cats" in TV studios, where none seemed especially comfortable, presumably because they were unavoidably well outside their usual familiar territories. I did know that cats were fantastic learners, despite their reputation for self-reliance and independence. The research showing how each and every cat learns how to use its meow (Chapter 4) had brought home to me, as much as anything could, how cats adapt their behavior to get by in the world that we expect them to live in. But unlike Sarah I had never put two and two together, that pet cats could—indeed should—be taught how to lead happier lives.

Everyone knows that an untrained dog is both a liability for its owner and a danger to itself (though there are several conflicting schools of thought as to how best to train a dog). I've never heard anyone complain about a cat being "untrained"—luckily for them, cats are far less of a social liability than dogs are. However, in some parts of the world, most notably in Australia and New Zealand, "dangerous cats" legislation is beginning to appear. Not of course for the same reasons behind "dangerous dog" laws—the danger is perceived as being toward wildlife rather than toward people. The science to support such legislation is somewhat lacking—for example, "cat curfews" have failed to halt the decline in some Australian marsupials—but its very existence demonstrates that for some people, cats are unwelcome additions to the local fauna, even though in many places they have been part of the landscape for hundreds, even thousands, of years, and the local wildlife seems to have adapted somewhat to their presence.

In many places (not just Australasia) the pressure is on to keep cats indoors 24/7. Some advocate this as a way of preventing them from hunting; others—including some rehoming charities—as a way of keeping cats away from motor vehicles, predators (for example, coyotes in North America) and aggressive neighborhood cats who wish to dispute your cat's right to roam anywhere, even out through his own cat flap. Despite these pressures, we're not in favor of turning cats into exclusively indoor pets. However, we acknowledge that some cats and other owners' situations do favor an indoor-only lifestyle, or at least require that serious consideration be given to the trade-offs between the negatives of outdoor access and those of lifelong confinement. Therefore, in this book we suggest a number of ways that training can provide possible solutions to these dilemmas—a cat can be called back just like a well-behaved dog when the owner senses danger that the cat cannot; diversity can be added to the indoor environment to satisfy the cat's need to explore and investigate; indoor (or indeed indoor/outdoor) cats can play games with their owners that mimic hunting and hence (hopefully) reduce the cat's instinctive desire to hunt; cats can be taught to play games with their owners that have little directly to do with hunting but simply enhance their relationship as well as providing both with amusement.

Ultimately, our ambition is to break down not just one but two preconceptions: first, that cats can't be trained; second, that cats can't benefit from being trained. We've always known that the first is demonstrably false. So far as the second is concerned, we believe that the well-being of the cats of the future depends upon a fundamental change in attitudes, a change that reflects current demands that all domestic animals should be "model citizens." The days when dogs were allowed to roam wherever they wished are long gone, at least in the West: for cats, a similar situation seems to be fast approaching. Not that we are advocating that cats should become just like dogs— the two animals are as unalike as chalk and cheese in terms of their basic natures and their fundamental requirements for a happy life, which can be (perhaps over-) simplified as "dogs need their people, cats need their space." The kind of training that we are advocating for cats is nothing like the "obedience" training that you'll find described in most dog-training books. It's much more about helping cats to adapt to the demands we increasingly place on them, demands that cats used to be expected to sort out for themselves.

Our hope is that if only cats could read, they would reward our efforts with whatever the feline equivalent of gratitude may be.

Conventions used in this book

If this were a science book, we'd be referring to the humans as "he" and "she," and the animals as "it." However, we quickly came to the conclusion that this convention was totally inappropriate for a book that aims to enhance the relationship between individual owners and their unique, individual cats: in particular, it felt completely wrong not to use the more personal "he" or "she" for the cats as well as the humans (and at least in the United States there are moves to adopt this new and more personal approach in scientific writing). However, we have no way of telling whether you, the reader, are trying to train a male cat or a female cat. So, to avoid clumsiness, throughout this book we will be referring to the cat as "he" (please don't be offended, lady cats, there is logic behind this, as you'll see). However, where we are referring to the cat as a species and not as an individual, we will refer to the cat as "it."

(continues)

Conventions used in this book *(continued)*

The eleven numbered chapters all follow the same format. Each begins with a general introduction to the way that cats perceive the world, relevant to the topic of that chapter, mainly written by John. Then follows the main body of the chapter, which describes how training can be used to address that topic and are written from Sarah's perspective (because she's the one with training experience). So wherever you read "I" or "me" or "my," that is Sarah. Owners are referred to throughout as female—sorry, that could be seen as a terrible cliché, but we're not trying to be sexist: rather, it keeps things simple and avoids references to cats and to owners getting muddled up. To male owners of female cats, especially, we offer our sincere apologies, and ask you to (hypothetically) switch genders while you read this book.

However, for some readers, the extent to which we personify cats in this book may not go far enough. Being British (Sarah is Scottish and John is English) we keep to our tradition of referring to cats as "pets" and their associated humans as their "owners." We do not follow the growing practice in the United States of referring to pets as "companions" and their owners as "guardians." For us, "ownership" of a cat implies responsibility and certainly does not imply the right to treat a cat as if it were the inanimate possession. We worry that the term "guardian" implies a legal status that arises from some mental deficiency in the animal—hardly an appropriate way to characterize the relationship between a cat and his significant person.[1] "Caregiver"—another suggestion we've seen—seems too impersonal, too transient, being much more appropriate for those devoted souls who look after cats in rehoming facilities. "Pet parent" is simply too anthropomorphic for us to stomach, especially as biologists who regard the term as typified by (though not restricted to) a genetic relationship between mother or father and offspring. So, for some readers we're going to be unashamedly non-PC—we can confidently say that we own our cats, because privately we can admit that, albeit in a slightly different sense, they own us!

Introduction

Why train a cat?
(and why cats aren't dogs,
and certainly not little people)

W HO ON EARTH TRAINS CATS? TRAINED LIONS AND TIGERS used to be a staple of circuses, until public opinion turned against them. Performing domestic cats seem more acceptable: Moscow has its Cat Theater, and the Amazing Acro-Cats tour the United States when they're not in demand for TV and film work. But why would anyone want to train their *pet* cat, except perhaps to show off their feline accomplice's talents to their friends?

This book has a more serious purpose: we aim to show you how training can improve not just your relationship with your cat but also your beloved pet's sense of well-being. That's not to say that the training won't be fun—it will, for both of you—but the distinction is that you will be producing a happy and well-disposed pet, not a circus star.

There are many everyday situations that our cats have to cope with as part of the deal of living with us. They don't easily digest the fact that humans come in many shapes and sizes, that men, women and children look and behave differently. Many find it difficult to adjust to living alongside dogs, or indeed other cats. They hate feeling trapped and don't understand that sometimes we have to restrict their movements for their own good, such as when we need to give

1

them medication. They don't like being taken to places they don't know or where they sense that there might just possibly be danger. While some individual cats seem to take some of these situations in their stride, most don't. By following the straightforward exercises described in this book, you will be able to give your cat a better life—and who would not want that for their treasured companion? Nowadays, we expect a great deal more from our cats than we ever used to, and training is the best way of helping your cat to cope with those demands.

Dog owners know that dogs can be trained, but the idea of training rarely crosses cat owners' minds. It is certainly true that an untrained dog can be a menace both to himself and the humans he lives with, whereas cats have gotten by for millennia without anyone deliberately attempting to train them. However, necessity is not the same as facility: just because few people bother to train their cats does not mean that shaping a cat's behavior is some kind of black art, suitable only for a select minority of professionals. On the contrary, every cat in the land can benefit from being taught how to cope better with tricky situations, such as accepting a pill or going into the cat carrier. And once the principle that cats do not think like dogs has been taken on board, training a cat is actually remarkably straightforward.

Fundamentally, the way that cats *learn* is very similar to the way that a dog—or indeed any mammal—learns, but cats have a unique way of analyzing and evaluating the world around them. To some extent this is due to the way that their senses enable them to perceive their surroundings, which, believe it or not, appear rather different from the version that we humans inhabit. However, it mainly comes down to the unusual way that cats both prioritize the information they are receiving and then react to it, both of which are somewhat unlike those of dogs, and even more different from our own. Much of what makes cats cats—their independence, their dislike of any kind of upheaval or social change, their fascination with hunting—makes perfect sense once their journey from wild predator to domestic pet is understood.

DOMESTIC CATS ARE TO BE FOUND IN EVERY CORNER OF THE globe. Worldwide, there are roughly three cats for every domestic dog, and although many of these cats do not have owners, in most countries pet cats are at least as popular as pet dogs. Yet the fact that some cats are owned while many are not hints at the possibility that as a species, cats are not yet completely domesticated. Indeed, cats do have a reputation for being rather independent animals, quite distinct from the much more emotionally needy dog. This is not to say that cats are cold and unemotional, as some people would have us believe, merely that they are less inclined to show their feelings at every available opportunity. And they are generally much easier to look after than dogs are, because they don't need to be walked and can cope with being left on their own for several hours at a time, a situation that many dogs find stressful (although few dog owners seem to realize this).

Ten thousand years ago there were no domestic cats, just thirty or so species of small wild cats, as well as a handful of big cat species, living in different parts of the world. All of these trace their ancestry back 10 million years to the first cat of all, known as *Pseudalurus*, from which all today's felids, from the lion down to the tiny black-footed cat, claim their descent. Fast-forward to about 2 million years ago, and we see the emergence of many of the types of wild cat that still inhabit the earth today. One group evolved in South America, including the ocelot, Geoffroy's cat, and the jaguarundi (which looks—and lives—more like an otter than a cat). Another group colonized central and southern Asia: among these were the shaggy manul, or Pallas's cat—which used to be considered a possible ancestor for the long-haired breeds of domestic cat until DNA testing ruled that out—and the Asian leopard cat, from which the modern Bengal "breed" is partly derived.[1]

Farther west, another set of cats evolved and began to spread into Europe. Among these was the ancestor of all of today's domestic cats, the wildcat *Felis silvestris*. This species occurs throughout Africa, western Asia and Europe—including the Highlands of Scotland, where the only British population of wildcat now teeters on the edge of extinction. The first reliable records of proper domesticated cats come from Egypt, about 6,000 years ago, but it is likely that the process of domestication had started several thousand years earlier, prompted by a key event in our own journey toward civilization, the emergence of house mice.[2]

The house mouse probably evolved when a new source of food appeared for the first time, the stores of harvested grains and nuts that our ancestors began to accumulate as they switched from nomadic hunting and gathering to settling in one place and stockpiling food to tide them over in lean times. Pottery had not yet been invented, and these stores, made from woven fibers, skins or unfired clay, would have been vulnerable to pests. Dogs had been domesticated several thousand years previously but seem to have been of little use in the war against mice and other rodents that feasted on the unprecedented concentrations of food provided by humankind's change in lifestyle. Into this scenario came wildcats, attracted by the new concentrations of rodents as inevitably as those rodents had been attracted by stores of grain and nuts.

The first civilization to be plagued by rodents were, we think, the Natufians, who lived to the east of the Mediterranean Sea, in an area overlapping what is now Lebanon, Israel, Palestine, Jordan and Syria, about 10,000 years ago. It is most likely in this region that wildcats first began to transform themselves into domestic cats, a theory supported by the DNA of today's pet cats, which is more similar to that of wildcats from the Middle East than it is to those that now live in Europe, India or southern Africa.

For hundreds, possibly thousands, of years, these cats would have only visited human habitations to hunt, retreating to the wild to rest and to raise their young—lives rather similar to those of today's urban foxes, except that the cats' efforts in keeping the rodents at bay almost certainly came to be much appreciated. Wildcat and domestic cat probably first diverged when a few bold individuals, more tolerant of humans than the rest, began to stay in the villages full time in between hunting forays. Possibly people encouraged this by providing secure places for them to sleep and to have their kittens. As generation succeeded generation, so the cats that were most tolerant of humans would have been able to spend the most time hunting, going about their business without being disturbed by our everyday activities in the way that most wild predators are. The undoubted appeal of kittens newly emerged from their nest would have led to these animals being handled, especially by the women and children, resulting in cats that were even more tolerant of people than their parents had been. Thus began the partnership between human and cat.

Yet even as these cats became tolerant of humankind, they would have still found it difficult to live side by side with other members of their own species. Wildcats are instinctively very territorial and aggressive toward one another. The males are intolerant of all other males and consort with females only once a year, for mating. Adult females are similarly aggressive toward one another, and although they are attentive mothers to their kittens for the first few months of their lives, they drive their offspring away as soon as they are mature enough to fend for themselves. As the size of human settlements grew, providing enough vermin to feed several cats year-round, so this territorial behavior would have become a problem, with cats distracted from hunting by constantly having to watch their own backs for attacks from potential rivals. Significant traces of such behavior remain today, as indicated by the difficulty that many cats have in sharing space with other cats that they have not grown up with.

Despite the limitations of their antisocial instincts, cats did then manage to evolve a rather limited form of cooperation—limited because it is restricted to females, while unneutered males remain fiercely independent (recall Kipling's "The Cat That Walked by Himself"—note the perceptive him). When there is enough food available, mother cats will allow their female kittens to remain with them even once they are old enough to breed themselves—and when they do breed, mother and adult daughters often place their kittens in a single nest and feed them all indiscriminately. Such behavior is now commonplace among free-ranging cats such as farm cats but has never been recorded in wildcats, suggesting that it evolved both during and as a consequence of domestication.[3]

Thus there are two key differences in behavior between wildcats and domestic cats. First, domestic cats can easily learn to be sociable with people, provided this begins during kittenhood. Wildcats, even those that are hand-reared away from their mothers, grow into wild animals that trust no one, except perhaps the person who raised them. Second, female domestic cats (and neutered males) can form friendships with other cats, especially, though not always, those that they have grown up with. Nevertheless, many pet cats remain intolerant of other cats for their whole lives, an enduring legacy of their wild origins, and the cause of much stress when they find themselves with ill-disposed feline neighbors.

Why was the wildcat the only cat to become domesticated? There were (and still are) several other species of cat living near the first permanent human settlements. These included the jungle cat, about the size of a spaniel, which the ancient Egyptians may have tried to domesticate. Certainly they kept them in captivity by the thousands, but they were probably too large to be effective in controlling mice and too dangerous to be allowed to roam freely where there were children (jungle cats are powerful enough to kill a young gazelle). Also nearby lived sand cats, smaller nocturnal animals with furry pads on their feet that enable them to hunt on hot sand and thereby live in desert areas that wildcats could not tolerate: however, the earliest peoples to store grains usually lived in wooded areas, typical wildcat habitat and probably too far from the nearest sand cat.

The transformation from pest controller to pet was probably a gradual one. The first evidence we have that cats were regarded as more than just exterminators comes from Egypt about 6,000 years ago.[4] We cannot be sure whether these cats had been imported from farther north or whether the Egyptians domesticated their local wildcats, but we do know that over the next 3,000 years cats gradually became more and more important to the Egyptians. Not just as pest controllers—though they became renowned for their ability to kill snakes as well as other vermin—but also as objects of worship.

Many different kinds of animals featured in ancient Egyptian cults and religions, especially big cats (lions and leopards), as well as birds such as the ibis. Domestic cats came to be specially associated with the goddess Bastet, whose original form, some 5,000 years ago, was that of a woman with a lion's head. Domestic cats were first depicted as her handmaidens, but by about five hundred years before the birth of Christ, Bastet had been transformed into something much more catlike, both in appearance and in character. Animal sacrifice was a big part of Egyptian religion at that time, and literally millions of domesticated cats were mummified and entombed as offerings to Bastet and other goddesses. Many of these cats were purpose-bred in catteries erected alongside the temples, but some of the mummified cats that have been recovered had been interred in elaborate decorated caskets and were evidently much-loved pets that had died of old age.

The ancient Egyptians' attitudes toward cats may seem rather strange to our modern sensibilities—some were sacrificed, some were revered, many must have simply earned their keep as humble pest controllers. Moreover, the whole history of the domestic cat from that day to this reflects shifts in the balance among these three conceptions. Although cats are no longer worshipped (in the religious sense) today, 2,000 years ago cat cults spread out of Egypt to all around the Mediterranean and persisted in rural areas right into the Middle Ages. The Roman Church's attempts to stamp out these and many other "heresies" had the unfortunate effect of sanctioning much cruelty toward perfectly innocent cats. Traces of these superstitions remain even today, such as the supposed association between black cats and witchcraft that we commemorate on Halloween, and events such as the annual Festival of Cats, held in the Belgian town of Ieper, which culminates with a basket full of cats being hurled from the top of the tallest tower in the town square—nowadays the basket is full of soft toys, but the use of live cats was discontinued less than two hundred years ago.

Many people find cats highly appealing, but a minority finds them repulsive, and over the centuries the predominant attitude seems to have fluctuated between these two extremes. However, the cat's usefulness as a rodent exterminator seems never to have been in doubt. For example, in tenth-century Welsh law, a cat was valued the same as a sheep, a goat or an untrained dog. Even then, cats seem to have been regarded as members of the family: the same law prescribed that in a divorce, the husband was allowed to take his favorite cat with him, but all the other cats in the household then belonged to the wife.

The emergence of the idea that cats could be pets first and foremost can be traced back to the eighteenth century, when cats began to be portrayed in purely affectionate terms. For example, the writer Samuel Johnson is said to have adored his cats Hodge and Lily, feeding them oysters and allowing them to scramble on to his shoulders. However, it was Queen Victoria who may have done more than any other person to popularize cats: her Angora cat, White Heather, was reputed to be one of the comforts of her old age and survived her to become the pet of her son, Edward VII.

As cats became more universally popular as pets, so distinct pedigrees emerged. Unlike dogs, where many breeds were originally intended for specific purposes, such as hunting, retrieving, herding and guarding, all pedigree cats are companion animals first and foremost. None is especially ancient: the DNA of Siamese cats shows that they only separated from their street cat cousins some 150 years ago, and today's Persians show no trace of their supposed origins in the Middle East. Pedigree cats do not, as yet, show the same degree of genetic problems as pedigree dogs do, and those problems that do exist are now being identified and steps are being taken to reduce and eventually eliminate them.[5]

More recently, successful attempts have been made to create novel types of cat by crossing domestic cats with wild cats of other species: these include the Bengal, derived from the Asian leopard cat; the Savannah, a cross with the African serval; and the Safari, derived from a South American species, Geoffroy's cat. These are really hybrids, even though they are often referred to as breeds, and their behavior can be as unpredictable and wild as their origins might suggest.

Today, the majority of cats are still nonpedigree, the product of thousands of years of natural selection, not deliberate breeding, and are therefore generally healthy and physically quite well suited to the environment they live in. They are nevertheless a highly specialized animal whose biology—not to mention psychology—needs to be properly understood if their well-being is to be protected.

LIKE DOGS AND INDEED HUMANS, CATS ARE MAMMALS, AND ALL three species share the same body plan. Reflecting their largely predatory lifestyle, dogs and cats have teeth quite unlike ours: they have prominent canine teeth, used in hunting, and their molars, which most mammals use for grinding, have been modified so that they act more like scissors when the cat or dog is chewing. Yet although cats and dogs have many similarities—for example, they are both meat-eaters—there are also many differences between them. Most of the time, cats keep their claws well protected in pockets of skin in their toes, extending them only when they want to use them. Dogs have immobile claws that wear down as they walk—their paws are designed for running and digging. And of course, cats are much more agile than are dogs. They lack collarbones, allowing them to place one front foot precisely in front of the other when walking along the top of a garden wall, and they use their tails like a tightrope-walker's pole to keep their balance. Thus, for cats, a home is a much more three-dimensional place than it is for a dog. Cats' ability to jump and climb means that they can make use of all of the space around them, not just indoors but also outdoors.[6]

The greatest differences among cat, dog and human lie not on the outside but beneath the skin. When it comes to choosing what to eat, dogs are much like us—we are both omnivores, able to live on a mixed diet of both animal and plant material and capable of existing on a vegetarian diet. Domestic cats, in common with the whole of the cat family, are strict carnivores. At some point during the course of their evolution, they became "locked in" to meat eating, through the loss of some of the key mechanisms that enable us—and dogs— to turn fruit, vegetables and grains into muscle and sinew.[7]

Hence until modern pet foods became available, cats were also "locked into" hunting as their only reliable year-round source of meat. This is part of the reason why cats continue to go out hunting even though we now feed them well—only a few generations ago, the ability to hunt was crucial for their survival. The other factor is that cats generally go for quite small prey. There are only about thirty calories in a mouse, so when they were still working for their living, cats had to kill about ten times each day just to survive. Such cats will hunt even when they've just been fed, just in case they don't find any prey for several hours: a cat that waited until it was hungry before setting off on a hunt would eventually starve.

Cats are often portrayed as "heartless killers," but a well-fed cat that goes out hunting is simply obeying instincts that served it well throughout its evolutionary history. Even worse for their reputation, pet cats often appear to "torture" or "play with" their prey, but this is an anthropomorphic interpretation. One explanation for this behavior is a sudden reduction in the cat's drive to hunt that happens just before or just after the kill, caused by the modern cat's excellent nutritional status. Or it could simply be that the pet cat has never learned how to hunt effectively. This also explains why many cats do not eat the prey they catch: to put it in a slightly more fanciful way, many cats lose interest in their prey at the moment when they recall that commercial cat food is tastier than mouse.

However, cats don't need to find real prey to satisfy their hunting instincts. It's not widely appreciated that when pet cats are "playing" with "toys," their behavior is so similar to actual hunting that they almost certainly "think" that that is what they are doing. Mouse-sized toys are manipulated just as if they were real mice; rat-sized toys are either avoided (not all cats are prepared to take on a real-life rat) or held at paws-length in the same way that a real live rat would be. Most tellingly, cats play more intently and intensely when they are waiting for a meal than when they've just eaten, reflecting the enhanced hunting instinct of a hungry cat. This overlap between play and hunting opens up the intriguing possibility that owners may be able to satisfy cats' predatory instincts simply by playing with them.[8]

It's unfair of owners to expect that an animal whose hunting abilities were prized until just a few generations ago should now abandon this habit. Nevertheless, most are perfectly reasonably revolted by the gory little "presents" that their cat leaves for them from time to time. And of course there is mounting pressure from wildlife enthusiasts for cat owners to curb their cats' attempts at predation—although the evidence points to feral cats, not pet cats, as being the real culprits. For the cat that persists in chasing after birds and mice, there are several devices designed to render them less effective as hunters, either by making them more conspicuous or by altering their ability to pounce. Most of these are worn around the neck, and although cats generally dislike them to begin with, this can be overcome using training.[9]

Cats, like all animals, gather information about the world around them—including their prey—through their senses, exquisitely tuned to their ancestral lifestyle as highly specialized hunters. Their hearing covers a much larger range than ours, and unlike us they can hear the high-pitched squeaks that rodents make to each other, well above our range of hearing, which is why we refer to them as "ultrasonic." The external parts of their ears (the pinnae) are not only highly mobile but can be moved independently of one another, enabling cats to pinpoint where sounds are coming from much more accurately than we can. Even the little corrugations inside their ears do more than just hold the pinnae upright: by subtly altering the pitch of a sound, they allow the cat to deduce the height it is coming from.

Cats' eyes are even more specialized for hunting. Compared to the size of the head, they are enormous—almost as big in real terms as our own. This enables cats to see where they are going on even the darkest nights, and their retinas are specialized for the same purpose, packed with receptors for nighttime vision, which like ours is in black and white. But unlike us, cats have very few daytime color receptors—they can distinguish among some colors but pay far less attention than we do to what color something is. Another specialization for nighttime vision is the reflective layer at the back of the eye, the tapetum, which inspired the "cat's-eyes" used in roads. Any light that misses the receptors on the way into the eye gets reflected back through the receptors: the small amount that misses the second time around is beamed out of the cat's eyes, giving them their characteristic green "eye shine."

Eyes that are so specialized for night vision can struggle somewhat on a bright sunny day. If the cat's pupil contracted to a point as ours does, the amount of light would probably hurt the cat's eyes, so instead their pupils close to a slit. Sometimes even this is not enough protection, and the cat will half-close its eyes so that only the middle of the slit is exposed (cats will also half-close their eyes when they're feeling particularly relaxed, whatever the level of light).

Such large eyes are also difficult to focus, and anything that is literally under the cat's nose will appear blurred. To compensate, cats are able to sweep their highly sensitive whiskers forward, substituting touch for close vision. The whiskers, and other less obvious tufts of sensory hairs on the head and the sides of the legs, also help the cat to feel its way through very dark places.

Although dogs are renowned for their sense of smell, it is less widely known that cats also have highly sensitive noses—perhaps ten times less sensitive than a dog's but at least a thousand times more sensitive than our own. Thus, cats, just as much as dogs, live in a world of odor that we can have only the vaguest idea about. Cats can certainly locate mice by the trails of scent that they have left moving through the grass, but perhaps more importantly for pet cats, they almost certainly pick up a great deal of scent information concerning the whereabouts and activities of other cats in their neighborhood. Although they do use their noses to do this, cats also analyze the odors of other cats using a secondary olfactory apparatus, known as the vomeronasal organ, which lies between their nostrils and the roof of their mouth. To bring this into play, they half-open their mouths and "taste" the air. Thus, when a cat seems to have gone into a brief trance with its mouth gaping open, it has probably just come across a scent mark left by another cat.[10]

Thus, apart from their acute night vision, cats' sensory abilities are more like a dog's than like our own. Likewise, their brains are also rather similar, constructed along a pattern that's common to all the members of the Carnivora, and crucially different from our own, primate, brain (see nearby figure). Cats' smaller bodies inevitably mean that their brains are lighter than our own, but their brains are also, at 0.9 percent of their body weight, relatively smaller than ours, at 2 percent. Much of our brain's extra tissue is contributed by the "thinking" part of the brain, the cerebral cortex, which wraps around much of the exterior and is highly folded. The cat's cerebral cortex is both smaller and much less folded than ours (although it is slightly more folded than that of the dog), suggesting that cats probably don't rely on conscious thought to anything like the same extent that we do. Conversely, cats' greater reliance on their sense(s) of smell is reflected in the prominence of that part of the brain devoted to processing olfactory information (see nearby figure). Rather than being at the front of the brain as in cats and dogs, as our own brains grow the olfactory region is pushed down to the underside of the brain by the dramatic growth of the cerebral cortex.

These differences between us and cats in the way our brains are constructed must inevitably reflect differences in the way we think, but so far science has not been able to entirely pin these down. We

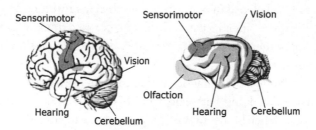

Side views of a human's brain and a cat's brain
(not to the same scale) showing some of the regions
devoted to the senses, to locomotion and to precise
coordination of movement (the cerebellum).

intuitively know what it's like to be a human, but it's much more difficult to perceive the world as it exists in a cat's mind. However, we can be virtually certain that our version of the world and the cat's version are different, and trying to understand what the differences are is essential for gaining an understanding of how cats react to our attempts at training.

One important question is, how do cats see us? The most widely accepted explanation for our huge cerebral cortex is that it enables us to comprehend social relationships in a much more sophisticated way than other mammals can. Lacking the necessary brain structures, cats must logically perceive their relationships with their owners (and with each other) in a far simpler way than we conceive of our relationships with them.[11]

One crucial difference between us may be what is often referred to as "theory of mind." When we talk to our cats, we can imagine them listening to us—and we know that they have minds of their own. Cats clearly identify the humans they know as individuals and react to what we do, but the scientific evidence indicates—and this can be hard for owners to grasp—that they do not have any comprehension that we are *thinking* about them. The ability to conceive of and predict what another animal or human might be thinking is probably confined, at least in land mammals, to the most evolutionarily advanced primates (the apes) and is of course much more fully developed in our own species than in any other. However much we owners might like to think that our cats think of us just as we think of them, their very different brains mean that they almost certainly don't.[12]

What this means in practice is that cats can pay a lot of attention to what we're doing but seem to have little idea of our thought processes. When someone finds a piece of meat missing from the kitchen worktop and comes to the conclusion that their cat has stolen it, it's natural for them to go looking for the cat to scold. We would rightly expect a child who had appropriated a biscuit from the kitchen a few minutes earlier to know precisely why he or she is being scolded, even before the verbal explanation. It's therefore natural to expect a cat to be able to make the same deduction. However, because cats don't even realize *that* we think, it's impossible for them to realize *what* we're thinking.

A second major difference between our brains and cats' is that cats seem to live largely in the present. They do of course have excellent memories—otherwise training would be impossible—but those memories probably only surface when triggered by something similar happening in the present: for example, a cat that catches sight of a black cat through the window may at that moment remember other encounters he has had with black cats. A few minutes after the black cat disappears, he will be thinking about something else: cats don't seem to be capable of recalling memories whenever they want to in the way that you or I can (or at least, believe we can). A cat that hears his owner's voice saying, "Come here, kitty," will instantly remember the previous occasions when he's responded by running toward his owner—and receiving a food treat for his trouble—and so will (if he's not too distracted by something else) do the same again.

For the same reason, it's highly unlikely that cats can actually be "devious" or "scheming," however much we may like to interpret their behavior that way. Not only do they live in the present, they don't seem capable either of reflecting on what's happened in the past, or, perhaps more importantly, of planning for the future.

AN ABILITY TO UNDERSTAND AND PREDICT A CAT'S FEELINGS IS critical for successful training, yet misunderstandings also abound concerning cats' emotional lives. Although little study has been made of cats' innermost feelings, it has recently become possible to observe how brain activity changes in different contexts, by training dogs and other mammals to lie still long enough to obtain an fMRI scan—and it should not be long before this is repeated on a cat.[13]

Such studies show that the mammalian brain, whether in a dog, a cat, or a mouse, generates a common repertoire of simple emotions—happiness, fear, anxiety, frustration—those that can also be referred to as "gut feelings." Ultimately, training works by altering the circumstances under which these emotions are evoked. The reward-based training that works best with cats aims to decrease negative feelings—such as fear, anxiety and frustration—and increase positive feelings such as joy and affection, by changing the cat's associations between these feelings and its day-to-day experiences.

Many owners also believe that cats are capable of experiencing much more complex emotions, including many of those that, at least for us humans, are conscious experiences. These include jealousy—probably experienced at some level by the more socially savvy dog but possibly not by cats—and pride, empathy and guilt, which are almost certainly beyond the mental capacities of dog or cat. An owner who punishes a cat under the mistaken belief that it is capable of feeling guilty about a mess made while she was out will be harming her relationship with the cat (and possibly making it more likely that the cat will make the same mess again). Cats live "in the moment" and are incapable of associating an action they have performed with its consequence—negative or positive—if that consequence occurs even a few minutes later, let alone an hour later. Rather, a cat will associate the punishment, or reward, with whatever is uppermost in its mind at the time. In the case of the cat that has made a mess while its owner is out, this is most likely to be the owner's recent return. Owners who pursue such tactics can be astonished when the cat no longer greets them warmly—and the messing is likely to get worse as the cat's general level of anxiety rises.

CATS AND DOGS DIFFER FUNDAMENTALLY IN THE WAY THEY interpret social information, whether it is coming from other members of their own species or from people. Cats are also rather different from the average dog in the way they react to things they haven't come across before.

There are many different ways to set about training a pet dog, but they all have two things in common: they rely on the dog's unique sensitivity to human body language, and also their innate affection for whoever looks after them. Dogs are fundamentally sociable, reflecting their origins as wolves living in cooperative family groups. Domestication has profoundly altered the dog's mind from that of its ancestor, the wolf, to the extent that dogs crave human attention—a dog that has been abandoned will attach itself to anyone who treats it kindly, even for as little as fifteen minutes. Experiments have shown that dogs are more attentive to human gestures than even chimpanzees, supposedly the most intelligent of all the animals, apart from ourselves. Both of these abilities have been an essential part of the relationship between dogs and humankind for thousands of years, enabling us to use dogs in a whole host of roles, including guarding, herding and hunting, although of course nowadays most dogs are kept simply for the companionship they so effectively provide. All of this means that training a dog will always be different from training any other animal, because they simply have a different perception of human beings.[14]

The profound differences in the way that cats see us compared to the way dogs do can obscure the fact that these two species actually learn, for all intents and purposes, in the same way. It's their *motivations* for learning that differ: *how* they learn—and how good they are at learning—are quite similar. Because dogs are thought of as trainable, and cats not, it can be easy to assume that dogs learn more than cats do. Although it's impossible for us to know what it's like to be a cat—or a dog—it's safe to say that both are excellent learners.

Cats, descended from solitary territorial animals, are generally much more wary of social contact, and many have to learn to trust other cats, and people, and often can only do this one individual at a time. Cats' primary attachment is to place, not people. That is not to say that they cannot become affectionate toward their owners, because they obviously can, but their affection can only blossom in an atmosphere of security. A cat's first priority is to find a safe place to live and a reliable source of food, which for pet cats is usually satisfied in their owner's house, before they can start to form strong social attachments.

For many people, cats, and especially kittens, have a heart-melting appeal that is not easy to explain fully. The appeal of kittens is now known to work at a fundamental level, triggering activity in some of the same parts of the brain that respond to the sight and sound of a human infant. The enormous popularity of kittens and cats on the Internet can almost certainly be attributed to this innate response of ours. However, this phenomenon alone cannot explain why so many of us go beyond their initial appeal to form lifelong attachments to our pet cats. Most cat owners spontaneously describe their pets as members of the family, and although science cannot yet entirely explain why this seems to come so naturally to so many people, it does guarantee that many cats are given the care and attention they need.[15]

Cats, however, do not automatically regard humans as their best friends. Worldwide, many cats stay wary of people for their whole lives. Cats (like dogs) need to start learning how to interact with people when they are very young. If kittens don't have at least a little friendly contact with humans during the crucial period of between two and eight weeks old, they usually go on to become feral cats. This remarkable flexibility in their social preferences shows just how malleable the feline brain can be—in contradiction of their reputation for being aloof and inflexible. Although they do become more set in their ways as they get older, cats retain the ability to assimilate new experiences and learn new responses throughout their lives.

The difference between cats and dogs in the strength of their attachments to people explains why traditional training methods based upon punishment are always counterproductive when applied to cats. Physical punishment is bad news for the welfare of dogs and cats alike, but dogs become so strongly bonded to their owners that their attachment can appear undiminished even when they are evidently aware that it is their owner who is causing them discomfort (as with a choke chain) or even pain. Just as abused children instinctively cling to their parents, so dogs continue to associate with people who have physically punished them (although at the same time their body language usually betrays telltale signs of the abusive element of the relationship).

Not so cats. Cats will naturally flee any situation that they find aversive, and so if they guess that these unpleasant feelings are in some way associated with their owner, their affection for their owner will instantly diminish. Even mild punishments that cause only minor discomfort and merely startle the cat can have this effect. For example, a water sprayer is often recommended as a way of deterring a cat from leaping onto the kitchen worktops: the sound of the spray, which resembles a cat's hiss, and the feel of water on the cat's skin are both mildly aversive. But ask yourself, What does the cat associate this with—the act of jumping onto the worktop, or the sight of the owner's hand? Only if the cat can be made totally unaware that the owner is involved will such a punishment achieve the desired result without damaging the relationship built up between the owner and her cat: even then, a cat that is already somewhat anxious may be made more so. The reward-based methods that we recommend in this book are vastly preferable, and there should be no need for any punishment any more severe than withholding the treat or toy that the cat desires.

Much abnormal and undesirable behavior in cats arises from situations where they feel their security is threatened, perhaps by a new cat being unceremoniously introduced to "their" home, an aggressive cat moving in next door or a new baby joining the household. Yet much of the stress that cats undoubtedly experience in such situations can be reduced, even eliminated entirely, by training. And training should also play a major role in preventing such problems from arising in the first place, for example, when used as part of the preparations for introducing a new cat to the home.

When it comes to encounters with the unfamiliar and the potentially threatening, cats are at a double disadvantage to dogs. As generally solitary animals, cats cannot depend upon strength in numbers, unlike the far more sociable dog. And, of course, cats are smaller and therefore more vulnerable than the average dog—and even more so than the dog's ancestor, the wolf, from which much of the dog's social self-confidence probably derives. Thus, most cats' standard reaction to anything unfamiliar is to keep their distance and at the first hint that trouble may be brewing, to run away. Some cats are temperamentally more timid than others, but it's a rare cat that is bold enough to stand his ground in the face of apparent adversity. Thus, very few cats ever learn much about how to deal with an unfamiliar situation, other than to put some distance between them and it. If such experiences are repeated, the negative connotations can only increase: just consider most cat's reluctance to enter their cat carrier, although they will actively seek out a similar-sized cardboard box in which to take a nap. For this reason, training often has to start with changes to the cat's environment that enhance its general feelings of security, giving the cat enough confidence to face its fears, albeit in a much more dilute form: only then can more positive connotations be built up.

EVEN THE MOST SECURE, CONFIDENT, AND WELL-SOCIALIZED CAT will find itself in situations that it deems unpleasant. Long-haired cats need frequent grooming, because their tongues have evolved to cope only with their ancestors' short coats. Cats' welfare has been improved immeasurably by the availability of cat-specialist veterinary surgeons, but try telling that to a cat that's resisting being put in the cat carrier with tooth and claw! Indeed, cats often miss veterinary appointments because their owners can't get them to the office, and much of the oral medication prescribed for cats never reaches its intended target because the cat spits it out or even refuses to have its mouth opened at all.

Cats, rather like small children, find it impossible to understand that something that's unpleasant in the moment will be good for them in the long run. Many long-haired cats experience discomfort from knots in their fur but equally don't like the process of being groomed. All cats instinctively balk at being thrust into a confined space, and traveling in a motor vehicle does not come naturally to them. We humans can tolerate the small indignities of a medical examination because we can comprehend that it's for our own benefit: unless taught otherwise, cats probably perceive a veterinary examination as the attentions of an unusual and definitely unwelcome kind of predator. Many cats are timid by nature, becoming nervous when unfamiliar people come into "their" house, and need to be provided with stage-managed opportunities to build up their confidence. Some cats can even be particularly sensitive to being picked up or stroked.

Dogs are usually so strongly attached to their owners that they are generally much more tolerant of small discomforts. Cats, much of whose sense of security comes from their physical surroundings, can be put off by even a small incident that they find particularly unpleasant. If the incident is repeated, cats may get into the habit of withdrawing whenever the same situation looks like it's going to occur. Every year, a significant number of cats "vote with their feet" and move to a nearby household they find less stressful. This shows just how effectively cats can learn about the consequences of interacting with the world around them. Rather than leaving a cat at the mercy of its instincts to learn lessons that served its ancestors well, surely it is better for the owner to teach lessons better suited to the complexities of life as a pet—in other words, use training to help the relationship with her cat run more smoothly.

Training can also be useful when introducing an additional cat into the household. Because cats are descended from solitary animals, their ability to get along with other cats is generally quite limited. Those free-ranging cats that coexist alongside others tend to be born into the group, rather than join it as an adult. Of course, no two cats are precisely the same, and they do vary in how accepting they are of members of their own species. Nevertheless, and contrary to many owners' preconceptions, introducing a resident cat to a new cat is as complicated as introducing two dogs to one another, and can often be more difficult. Owners who get this wrong can find that they have inadvertently created a lifetime of anxiety and stress for both their cats: along with an understanding of how each cat may feel threatened by the other, training can play a big part in making such introductions run smoothly.

ONCE ITS BASIC NEEDS FOR FOOD, WATER, A CLEAN LITTER BOX and a safe place to sleep are met, a cat's happiness usually revolves around how secure it feels. Cats do not mind being left on their own (unlike dogs), but they do value a predictable routine and stability in their social environment.

One issue that vexes many cat owners is whether to allow their cat outdoors. There are essentially three benefits from keeping a cat indoors: it will not be able to disturb or kill wildlife; it is not exposed to road traffic or people who wish to harm cats; it cannot get into disputes with other cats in the neighborhood. However, the wholly indoor cat is a comparatively new concept, and pet cats, whether indoor or outdoor, are descended from cats that hunted for a living and maintained extensive territories out of doors. In evolutionary terms, it is far too soon for cats to have lost their drive to range and explore. Hence, there is a significant risk that a cat kept indoors will suffer from frustration or boredom. Owners who decide that it is too dangerous for their cat to be let out do need to take careful steps to prevent this, and training can play a major part. They should also construct the cat's environment in such a way as to provide as much appropriate stimulation as possible, for example, by allowing access to an enclosed outdoor run, playing with the cat several times each day and using prey-like toys to satisfy its hunting instincts.

Cats that are allowed outdoors have more options than indoor-only cats, and although they will undoubtedly find this stimulating, some cats are unfortunate enough to find their right to go out challenged by another cat—which may even be so bold as to actually enter the house, adding to the resident cat's distress. Even if there is little direct evidence of such a conflict, the cat may show signs of being under stress, including spending an excessive amount of time hiding, urinating and defecating in the house outside the litter box, and even expressing its inner tension by attacking its owner.[16]

SOME SITUATIONS THAT PET CATS FIND THEMSELVES IN ARE A problem only for them; others are also a problem for their owner. Either way, training can often provide a solution that benefits both parties. As we see it, the primary purpose in training a pet cat should always be improving the cat's own sense of well-being, although owners will also find that at the same time they will reap the considerable reward of having a happier, easier to manage cat.

A cat will do something only if it feels like it. Our task ahead will be to ensure that the cat wants to do what is in its own best interest, even when its instincts tell it otherwise.

CHAPTER 1

How Cats Learn
And what you can do
to make it easy for them

T RAINING A CAT IS NOT A MYSTERIOUS PROCESS, BUT IT WILL
make sense only with an understanding of how cats learn. Cats
may have a reputation for inscrutability, but they are actually very
adaptable animals, and they learn in just the same way that all mammals do—including dogs. Cats are just as smart as dogs; it's just that
they have their own motivations and priorities, and because these
are less widely appreciated than those of dogs, cats have acquired a
reputation for being untrainable. Nothing could be further from the
truth: cat owners can easily acquire the skills and knowledge they
need to change their cat's behavior, and not just for their own benefit but for their cat's, too. Cats can be taught that commonplace situations that they instinctively dislike, such as being brushed or
being put into the cat carrier, can in fact be pleasurable experiences
rather than something to fear. Moreover, an appreciation of how
cats learn is also key to understanding much of their everyday behavior, for cats are much more responsive than their reputation for
stubbornness and independence might suggest.

MANY OWNERS SEEM UNAWARE THAT THEIR CATS ARE LEARNING
all the time. When we asked Mrs. Smith if she thought that Smoky,
her cat, had learned much in the past week, she replied. . . .

Smoky hasn't really learned anything recently as she has spent most of the week indoors. She usually likes pottering around outside when I am out at work but she hates getting wet and we have had an awful few days of rain. She is a really sweet affectionate cat and we often enjoy cuddles on my lap but to be honest, this week, she has been a bit of a pest. I have had the week off work and have been trying to get some baking done while I cannot get out in the garden. Smoky has, however, had other ideas and has been jumping up on the kitchen worktop as I have been trying to bake. I kept telling her that she was a silly girl as I lifted her off the worktop and gave her a quick but affectionate fuss. Even when the cake was in the oven and I went to check my e-mails, she kept getting in the way and trying to sit on the laptop. You would think that the fact she got some fuss from me would have satisfied her and let me get on with my jobs, but it didn't. In the end, on both occasions, I ended up feeding her early so I could get on with what I was trying to do. And even though the weather has brightened up today, Smoky has decided to stay in and pester me, rather than go outside and get some fresh air. She is driving me to despair.

What Mrs. Smith seemed not to realize was not only how much Smoky had learned on those few rainy days but also just how much she had unintentionally influenced that learning.

First, we can tell from Smoky's behavior that she had already learned that going outside when it is raining makes her cold and wet, and so it is better to stay inside where it is dry and warm. She had also learned that her owner does not go off to work every day but sometimes stays at home, and, as she is the kind of cat that enjoys her owner's company, had found that staying inside in the daytime would give her this reward. However, Smoky had also observed that Mrs. Smith sometimes did things that did not involve her, and as a consequence, she was not always paying her the attention she desired. By placing herself in between Mrs. Smith and the task she was focused on—the kitchen worktop or the laptop—Smoky got some attention. Mrs. Smith thought that Smoky understood when she told her not to interfere, but in Smoky's mind, being picked up and talked to was positive attention. It did not matter that Mrs. Smith was telling Smoky she was a "naughty girl": all Smoky understood was the tone of voice—which was gentle—and that jumping up resulted in a cuddle. Even better, when Smoky did this several times, she got fed quicker—the jackpot! Having learned that getting in between Mrs. Smith and whatever task she was doing resulted in attention . . . and . . . (bonus!) . . . food, Smoky was bound to try the very same behavior again the next day. Inadvertently, Mrs. Smith had taught Smoky that behaving in this way leads to good outcomes (generally getting in the way produces food!). If Mrs. Smith had fully understood how Smoky learns, she might have gone about her actions slightly differently, achieving some peaceful baking and laptop time while still keeping Smoky happy.

Cats learn all the time, regardless of whether we are intentionally trying to teach them something. Some associations are learned after just a single exposure, especially if the outcome is unpleasant: for example, the cat that strays into a neighbor's garden and gets chased by their dog will have immediately learned never to enter that garden again, or at least never when it can see that the dog is there. Other events need several repetitions to consolidate the learning to the point where the cat reliably changes his behavior in those circumstances; for example, a cat may need to receive food treats several times from a new visitor to the home before changing the way he behaves toward this person.

Different experiences will have differing outcomes for the cat. Some will be consistently positive, others consistently negative and many will be neutral (i.e., either the cat hasn't noticed the outcomes or has found them of no concern either way). Further experiences can either reinforce what has previously been learned in that particular situation (if the outcome remains the same) or start to teach something new (if the outcome changes—for example, from something nice to something less nice). All these experiences and outcomes are processed in the brain where learning takes place, memories are laid down and emotions and feelings arise. These in turn will all influence how the cat behaves, not just at the time but also in the future.

THE SIMPLEST TYPE OF LEARNING—COMMON TO ALL ANIMALS that possess a nervous system, from worms to humans, and so simple that it's debatable whether it can really be called learning—is known as *habituation*. This is a way in which animals learn to ignore those parts of their environment that have no special consequences and are therefore irrelevant. Upon repeated exposure to such things or events, cats learn to perceive them as harmless and so ignore them. For an animal as well endowed with sense organs as the cat is, continually focusing on the irrelevant diverts vital attention and energy away from events that may have an impact on survival—for example, nearby prey or predators. Thus, habituation is a vital learning process. For example, a kitten that arrives in a new home may startle when it first hears the ring of its new owner's mobile phone. However, after several repetitions of that phone ringing, the kitten will have learned that nothing happens at this time of any relevance and therefore will stop reacting to the ringing sound. In technical terms, the kitten has *habituated* to the sound of the mobile phone. Habituation is therefore an important process of learning, not just for kittens, but also for adult cats experiencing new environments such as moving to a new home. Cats, just like us, have to learn what is important and what is not, if only to avoid sensory overload.

Sensitization is the opposite of habituation. In this process, repeated exposure to an event leads to an increased reaction from the animal, as opposed to the reduction in response and eventual ignoring that characterizes habituation. The crucial difference is that now, the repeated exposure is to something that the cat instinctively dislikes. For example, several visits to the vet's that include unpleasant experiences, such as an injection, can lead to the cat becoming fearful of the vet, even on subsequent visits when the vet is friendly to the cat and no injection is planned. Also once a cat has become sensitized to one situation, it may show the same reaction in other similar circumstances. For example, the same cat may become wary or fearful of new people who simply look, sound or even smell like the vet. He may even become fearful of new environments that remind him of the trip to the vet. Sensitization is a powerful protective mechanism that helps cats avoid anything they perceive as potentially dangerous. One of our goals in training cats will be to teach them that encounters with vets—and many other situations—do not need to be perceived as potentially dangerous, thereby preventing sensitization before it has a chance to occur.

Both habituation and sensitization change the strength of the cat's existing reactions, but they don't help the cat to develop any new responses: for this, more complex learning processes are required. The most straightforward of these, known as *classical conditioning*, occurs when a cat finds out that some specific event reliably predicts that something else is about to happen. When a cat meows and runs to his owner as soon as he hears the cupboard that contains the cat food opening, he is responding to classical conditioning. The cat has learned that the sound of that cupboard door (in itself, a meaningless sound) predicts that food is on its way. Several repetitions of hearing the cupboard open just before the food arrives are needed for the cat to learn the predictive value of the sound. Once such learning has taken place, that specific sound elicits positive feelings in the cat's mind similar to those triggered by smelling or tasting the food. The cat does not have to learn that tasty food makes him feel good; this is an involuntary response, built into the cat. What the cat does learn is that things other than the taste and sight of food can create such feelings: in this case, the sound of the cupboard door. Such a learning process relies on a consistent pairing—the sound of the cupboard door always being followed quickly by the presentation of food. To begin with, the cat may make mistakes, such as responding to the sound of *any* kitchen cupboard opening, but most are then able to refine what they know, by learning that only the distinctive sound of *that* cupboard is reliably followed by the appearance of food.

CLASSICAL CONDITIONING HELPS A CAT TO MAKE BETTER SENSE OF
its environment, but a different kind of learning, *operant conditioning*, is needed for the cat's behavior to change. Operant conditioning involves the consequence of a cat's own actions influencing how
it feels and, thus, what behavior it should feel like performing next.
The consequences that result from any behavior can be classified
into four different types (see nearby box).[1]

Operant conditioning explains why Mrs. Smith's behavior caused
her cat to jump up and impede what she was doing, not just once
but several times. The positive feelings caused by the stroking, gentle talking and being fed encouraged her cat to repeat the behavior
of jumping up and interfering with what she was doing (i.e., Consequence 1). More formally, we would say that the behavior has
been *reinforced*. By this, we mean that the cat has found that the
behavior has a rewarding outcome, and, thus, he will be more likely
to perform the behavior again, in an attempt to recreate the positive
outcome.

For a cat to learn that any outcome is genuinely associated with
his behavior, it is usually essential that the consequences (positive
or negative) occur immediately. If not, the cat is unlikely to make
the connection. However, in some instances classical conditioning
can bridge the gap. For example, a cat may wander from the garden
into the kitchen. The owner would like to reward this behavior (so
that it occurs again) by giving the cat a food treat, but she may have
none to hand at that precise moment. However, if the cat has already learned, through classical conditioning, that the sound and
sight of the cat treat tin being touched is followed by a food reward,
then simply by reaching for the (empty) cat treat tin at the moment
the cat comes in from the garden, the owner will be able to "buy"
herself some time to reward the cat with food: the action of reaching
for the treat tin tells the cat that the real reward is on its way.

Cats do not respond well to anything nasty happening (unfortunately for them, dogs are much more tolerant in this respect). It is
really important to be mindful of cats' natural tendency to withdraw

The four types of consequence that trigger operant conditioning[2]

Scenario: Your cat sits on a laminated floor in front of you. He suddenly jumps up on to your lap.

Consequence 1:
    Something *good* is *presented* (e.g., you give the cat a food treat)

Consequence 2:
    Something *good* can *end* or be *taken away* (e.g., you stop feeding the cat treats and ignore the cat)

Consequence 3:
    Something *bad* can *start* or *occur* (e.g., you stand up and walk away—or you push the cat back onto the floor)

Consequence 4:
    Something *bad* can *end* or be *taken away* (e.g., while he is on your lap, he is off the cold floor)

from the slightest sign of trouble. Although learning will undoubtedly occur as a result of something negative, and especially any sort of physical punishment, use of such punishment can have a disastrous effect on the cat-owner relationship. A cat that has been physically punished is highly likely to respond in one or even several negative ways. First, it may become fearful of its owner, and even become fearful of other people, through sensitization. The cat's fear can be expressed as an aggressive response directed at whoever originally delivered the physical punishment or indeed, at anyone nearby. Fear can also cause a cat to try to escape or avoid further interaction of any kind. Additionally, the use of punishment generally causes a reduction in any kind of spontaneous behavior from the cat in the presence of the owner, thus making future training more difficult. Finally, punishment may tell the cat what not to do, but it doesn't help him to learn what the right thing might be. Moreover, all of these outcomes are distressing for the cat and likely to have a detrimental effect on his quality of life. For cats, successful training relies on rewarding the desired behavior and ignoring unwanted behavior. Keeping this approach at the center of all training should ensure a positive relationship, as well as a successful and happy learning experience for both parties.

Although cats learn a great deal from their owners, they can also learn from other cats that they get along well with. Kittens naturally learn a lot from their mothers. Both kittens and adult cats will learn to perform a task quickly, after simply watching an experienced cat complete the same task. Cats that live together are often believed by their owners to have "taught" one another particular behaviors—for example, how to use the cat flap. It's not clear whether the second cat actually learns how to perform the behavior directly from the other cat or whether the more skilled cat's actions simply draw the other cat's attention to the cat flap as something worth investigating. It's also not known whether cats are capable of actually imitating our actions (probably not), but we can easily use their natural curiosity to draw their attention to those features of their environment that we want them to learn about. Then, by providing the appropriate consequences, owners can make sure that desirable behavior occurs again and unwanted behavior does not.

Cats learn spontaneously, all the time, mostly by discovering reliable associations between events or features in their environment,
just as Smoky did. Thus, it is good practice to start observing your
cat, paying attention to his body language after he performs certain
actions and noticing whether you see the same behavior occurring
over and over again. For example, can you decipher whether a particular action your cat performed led to a positive, neutral or negative outcome for your cat? Can you start to see patterns in your cat's
behavior? What do you think might have been the cause of your
cat's little quirks and idiosyncrasies?

CATS LEARN ALL THE TIME AS THEY GO ABOUT THEIR EVERYDAY
lives, but we can boost their chances of learning what *we* want to
teach them by ensuring that they are in the right frame of mind. Just
like us, cats learn best when they're comfortable and free from distractions. They are naturally sensitive creatures that flee from any
threat or uncertainty; thus, the best place to teach a cat is somewhere he finds quiet and familiar. Just like people, cats need to be
free from distraction if they are to learn effectively. Although most
of us find it hard to ignore a ringing telephone, cats, with their acute
senses of smell and hearing, can be distracted by things we barely
notice. For example, both owner and cat may find it hard to focus
when the noise of the washing machine indicates that it is on full
spin cycle. However, the average person would barely notice the
faint odor of a piece of frozen meat defrosting on the kitchen worktop, although it can be overwhelmingly enticing for a cat, tempting
him to jump up on the worktop and investigate (cats are opportunists, and the chance of a free meal is something they rarely pass up).
For a cat, distractions do not include only sounds that we might
deem to be loud, irritating or unexpected but also enticing or unfamiliar smells and sights. For example, training in front of a window
that looks out onto bird feeders may appear to be no problem to
you, but the sight of birds fluttering around may engulf your cat's

attention, particularly if he is partial to hunting. Thus, before start-ing any training, it is important to think of what may be distracting from the cat's perspective.

Also, just like us, cats learn best when they feel comfortable: not too thirsty, not too hot or cold, not too tired nor in need of relieving their bladder or bowels. Thus, when selecting where in your home to begin training your cat, make sure there is fresh water and a litter box available (or access to the outdoors if your cat does not use a litter box). The temperature should be comfortable, and your cat should have the opportunity to retreat or rest if he so desires. Her-bie, for example, found it hard to concentrate on any training task when the log burner was well stoked; the warmth was just too irre-sistible. As an Asian (a breed that has a single coat, as opposed to the traditional double coat of the domestic short hair), Herbie was always seeking out a heat source. It was never long before he opted out of training in favor of lying full stretch, fast asleep in front of the stove when it was lit. Thus, I always carried out his training sessions before the stove was lit and we settled down for the night.

Cats don't usually learn well immediately after they've eaten: a certain degree of hunger is needed before a food treat becomes re-warding. A food treat delivered immediately after a specific behav-ior acts as a positive outcome for a hungry cat, encouraging the cat to perform that behavior again. Thus, it is important that your cat is feeling hungry so that he is motivated to engage with you for a food treat. However, how hungry your cat is can influence how well he learns, and this in turn depends very much on his personality. Being too hungry can inhibit training, as the cat may be more focused on the food itself than on learning which specific actions of his are be-ing rewarded.

Once the environment is set for minimal distraction and maxi-mum comfort, the next stage to consider is the construction of the teaching toolbox. And not just metaphorically: it is literally a good idea to have some form of robust box in which to keep most of the training aids. Having everything in one handy place makes it possi-ble to do a few minutes of training here and there, and to take advantage of teaching opportunities whenever they arise. Because cats spend a great deal of the day asleep, it is important to be ready to seize the moment for a short training session whenever you find they are awake and alert. Luckily, cats learn best in short bursts, and so taking advantage of these small windows of opportunity will naturally lead to the greatest success. As the cat learns that training is part of his daily routine, you may find that when you are around, he spends more time awake: after all, he now has a new, exciting, stimulating and engaging pastime to share with you.

By having all your training essentials in one portable container, the box itself can act as a signal to your cat that you are going to do some training. Very quickly, your cat will learn that when the box comes into view, he will soon have access to the exciting rewards inside (in itself, an example of classical conditioning), and, as a consequence, he will be engaged to work with you.

The rewards are the most important items in the training toolbox: successful teaching of a cat is ultimately based on being readily equipped to reward the behavior that you desire with something the cat really values. Rewards can take many forms. Cats, just like people and dogs, often become disinterested if they receive the same reward over and over again; therefore, it really is important to have a variety to hand so that the reward can be changed before the cat has had enough of any one type. This exact point is superbly illustrated by a study we conducted in which cats were repeatedly offered a toy to play with. When the toy presented was the same each time, the cats reduced the amount they played with it to almost nothing, showing they had become habituated. However, when the toy was changed for a different one each time, the cats continued to play, illustrating they were not "bored" with playing; they simply needed a fresh stimulus to remain motivated.[3]

At the beginning of training, the most important rewards are those known in the training world as *primary reinforcers*. These are things that cats instinctively find rewarding, and the most universal example is food. Being a carnivore, rewards comprised entirely of animal protein will be valued most highly by your cat. An ideal reward therefore consists of a tiny piece of cooked meat or fish (such as a quarter of a cooked prawn). Rewards with a high proportion of animal protein may also be useful, for example, a very small portion of the cat's normal diet (a single biscuit or a single meat chunk from a foil pouch or tin) or shop-bought cat treats—semimoist or air-dried meat snacks are often preferred. If part of the cat's daily food ration (treats or regular diet) is reserved for training rewards, training your cat using food should not result in him putting on weight. Part of his normal allowance can be weighed out or counted out for use in training and his regular feeds reduced correspondingly.

When teaching your cat something new, the rewards should come little and often. It is important that the rewards are small; first, so that they can be eaten quickly to allow you to speedily take up training where you left off before delivery of the reward, to keep learning momentum (cats generally eat more slowly than dogs), and second, to stop your cat from getting full too quickly. Cats generally prefer to eat small amounts often—recall that a free-ranging cat may eat ten small meals a day. As a guide, a single reward should be approximately half the size of your smallest fingernail.[4]

Cats, as a species, are often considered to be fussy when it comes to food, unlike dogs, many of which will eat anything and everything. Therefore, before starting training it is important to assess which food types your cat likes and just how much he likes them. There are lots of ways this can be done, from simply placing a selection of small pieces of food in front of your cat and seeing which he chooses to eat first, to trying different food types in puzzle feeders and seeing which ones motivate your cat to work at getting them out.

In fact, puzzle feeders are a great way to gear your cat's brain for some human-led teaching. By extracting food from a specially made device, such as a food-dispensing ball or maze, your cat inevitably learns that it is his behavior that has resulted in the reward appearing, whether that behavior is rolling a ball or pulling the treat through a maze until he can reach it with his mouth. This learning comes about in three stages: identifying that there is food in the device, usually by smelling it or seeing or hearing you place it in there;

the desire to obtain it; and finally, trying different ways to access the food. As your cat successfully obtains pieces of the food, he will learn that the action that immediately preceded this is the one he needs to try again in order to get more food. The same process occurs during more formal training, the only differences being that the trainer is the dispenser of food and that the way the food is presented can be adapted to help the cat learn the correct behavior more quickly than would happen through simple trial and error. Obtaining food rewards through puzzle feeders does involve considerable effort on the cat's part, but evolution has designed cats for this, because even greater effort is needed when hunting for food in the wild. Thus, the process of working for food is intrinsically rewarding for a cat and thus something we can encourage through formalized training. Puzzle feeders help your cat learn that his behavior can have rewarding consequences, "switching on" those parts of the brain devoted to making new connections.

A list of your cat's favorite treats in order of preference (there may be joint favorites) will be really useful when you come to start training. It means you can pick and choose the right treats to use depending on his mood and motivation at the time of training, and on the difficulty of the task in hand. For example, a hungry motivated cat who is really engaged and is learning a simple task can be rewarded with a few of the lower-value food treats, such as his usual biscuits. Those will be rewarding enough to keep him motivated to the task but not so tasty that he becomes overexcited and tries to swipe the food out of your hand. Likewise, if the cat is showing signs of disinterest in a task, the value of the reward can be increased to enhance his motivation. It is always a good idea to make sure that some of these food rewards are only ever given to your cat during training sessions. This will greatly enhance their value as a reinforcer of the desired behavior. Just like people, not all cats feel the same way about food; some get enthusiastic about it and will meow repeatedly at feeding time, others are content to just pick at their food now and then. For the latter, food treats of very high value (cooked meat and fish) are most likely to be rewarding.

There are many types of rewards other than food, even for cats who are highly food motivated. As well as eating (and sleeping), one thing cats particularly enjoy is playing. Because play is so closely linked to hunting behavior, it is a highly rewarding experience for cats (and scientists are increasingly using it to assess how happy animals are). Thus, for cats—particularly cats that are young, cats without access to the outdoors, and cats that are naturally playful—the opportunity for a game can be just as powerful a reward as food. The toolbox should therefore also include a selection of toys. The ideal toys for training sessions are ones that can be moved around quickly, those that generate short, intense bursts of play, and those that can be easily removed from the cat without risk of you getting scratched or bitten. Thus, wand toys (also known as fishing rod toys) are ideal. These comprise a small toy attached by string or elastic to a wand (which you hold). The wand keeps your hands well out of reach of the game while allowing you to move the toy quickly in straight lines. Such movement is a powerful way of eliciting play: wand toys mimic the movement of prey along the ground.[5] Others allow you to imitate the movement of prey items that fly: instead of a wand with string or elastic attached to the toy, they often have a wire that allows the toy to be flicked through the air.

Wand-based toys are designed to be used in interactive play and therefore should not be left out unattended. Keep a few of these types of toys solely for training sessions, storing them in your teaching toolbox when not in use—this will help them remain exciting, their availability clearly linked to training, thus enhancing their value as a reward.

Stroking can also be used as a training reward for any cat who is particularly affectionate and tactile. Most cats prefer to be stroked briefly and often, rather than for extended periods. In one of our studies that examined where on the body cats most enjoy being stroked, it was found that stroking concentrated on the top of the head and the face produced the most positive response from the cats. These areas are full of scent-rich glands that deposit communicative chemicals when cats rub their faces on objects. Cats that are closely

Herbie enjoys the reward of being stroked on his cheek.

bonded to each other will also rub their faces against one another during friendly greetings, and the same behavior is often directed toward human hands and legs. Such friendly behavior can therefore be mimicked by stroking these areas. However, because such behavior is reserved for only the closest companions, this interaction is likely to be rewarding only for those cats that both enjoy physical interaction and feel very comfortable with their owner.[6]

Cosmos spends a great deal of time rubbing his face against objects such as furniture and doors. Often he purrs when he performs this behavior, and if I outstretch a hand to him, he will rub his cheeks and chin against it. This is obviously something he finds enjoyable. He also loves to be groomed in these facial gland regions: while I'm grooming him, he will move himself into a position where he gets the most attention directed to his face. Taking this into account, during one training session where I was teaching Cosmos to sit, I decided to try allowing him to rub his face on a grooming brush, as a reward. We hit success, and the grooming brush is now a staple tool in our training box.

Cosmos standing before
the "sit" cue is given.

Cosmos sitting after
responding to a hand cue.

Cosmos enjoys rubbing his face
against a brush as a reward for sitting.

Understanding how cats learn will give you a solid foundation for comprehension of the various training exercises outlined in this book. However, not all cats learn in quite the same way: identifying your cat's individual idiosyncrasies will help you tailor each training exercise to your cat's unique likes and needs, and will set you up for the greatest training success.

CHAPTER 2

Understanding Your
Cat's Training Requirements

C ATS ARE UNMISTAKABLY CATS, BUT EACH AND EVERY ONE IS
also unique, as distinctive to its owner as one person is to another. Such uniqueness stems not only from physical characteristics
such as age, sex, health and breed but also from their individual personalities, moods and past and present life experiences, including
their current home situation. In combination, all these factors come
together in a way that makes each cat respond to training slightly
differently and that makes some cats slightly easier to train than
others. As a result, the way you train will need to be tailored to your
individual cat. For example, factors such as the best type of reward
to use, the optimal length of a single training session, the length of
time it will take for a cat to accomplish a training task and the best
training aids to use will all differ among cats. Taking some time to
think about your cat's unique characteristics as well as his current
mood will therefore be a useful preliminary, before any formal training is attempted.

Perhaps the first factor to consider prior to training is how old
your cat is. Thanks to much-improved veterinary care, cats can now
live on into a ripe old age, and there's no reason why an old cat
can't be taught *some* new tricks. However, like most animals—
including ourselves—cats learn fastest when they're young and
a little more slowly once they're older. Many senior cats (eleven

years and upwards) will still cope well with training but are likely to need more time spent on each element of training to accommodate their slower learning speed. This is particularly true for older cats who have had no training experience when younger. Although training may need more repetition for an older cat, each individual training session will need to be shorter, as older brains (and bodies) fatigue quickly.

Young kittens learn something almost every waking minute, but up until they are about ten weeks old, their brains are growing rapidly, and their concentration is extremely short. Any formal training sessions should be kept simple and last only a few minutes. For example, you could reward your kitten for choosing to play with a toy rather than with your hands or feet. Play is automatically rewarding, particularly at this age, and so a game with the toy becomes the reward for opting not to pounce on nearby humans. This can be done by simply removing yourself from the kitten's reach if he makes attempts to play with your hands or feet. Training sessions, even though short, will be most beneficial when frequent, up to three or four per day. These should, however, be well separated to allow the kitten time for rest and sleep in between.

Adaptations to equipment may be needed to make sure no physical injury occurs when training kittens that are still developing physically as well as mentally. The same applies for older cats that are less agile and mobile. For example, raised platforms may need to be lowered and cat flaps may need steps added on either side to prevent too large a jump in or out, whereas barriers such as baby gates may need mesh added to prevent a kitten from pushing through the narrow bars. Blankets and bedding used for senior and geriatric cats should be of a fabric that the cat is unlikely to get a claw trapped in—for example, fleece. Many older cats find it difficult to retract their claws, increasing the likelihood that they will get caught up in fabrics such as toweling and loose-knit wool.

Although it's sometimes said that female cats are more affectionate than are males, this probably harks back to the days before neutering was the norm for pet cats. Tomcats—sexually intact males—become more and more independent of people as they grow into adulthood, at which point they become obsessed with searching for receptive females and defending their territories against rival toms. Nowadays, the majority of male cats are neutered before they are six months old—indeed, owners who forgo this can find that their beloved cat has disappeared forever, chased away by the mature males in the neighborhood who have begun to perceive him as a rival. Female cats, if left sexually intact, generally stay close to home, at least until they find themselves coming into season. Then, the entire focus of their lives shifts, for a week or so, to the business of finding the best father for their kittens. Once neutered or spayed, a cat's personality seems largely unaffected by whether it is male or female and should therefore not impact greatly on training.

Unlike dogs, cats have evolved from a solitary ancestor and, as a result, tend to hide illness or injury exceptionally well. It is not advisable to draw the attention of potential competitors to the fact that you are incapacitated and unable to defend your territory. As with all animals, cats learn best when they're feeling fit and healthy, so it is important to pay close attention to your cat's health. Avoid training your cat if he shows any signs of illness or injury. Although long-term health issues that your cat always lives with may not preclude training entirely, they may need addressing within your training plans: for example, only certain types or amounts of food rewards can be used for a cat with dietary sensitivities or with diabetes. If your cat is on any medication or you are at all unsure how your training plans could influence your cat's current health conditions, please seek veterinary advice before commencing any training. For example, check with the vet before changing the timing of feeding or the food the cat receives. Any sensory impairments are also likely to influence your cat's ability to learn. For example, there is no use in verbally rewarding a deaf cat—but a low-level flash of a flashlight could be paired with a food reward through classical conditioning to teach the cat a food treat is on its way. For a cat with impaired vision, delivery of a food reward may need to involve strong-smelling food delivered within easy sniffing distance of the cat.

Only a minority of pet cats can claim to be of a particular breed, but some breeds do behave in rather characteristic ways, and this can be part of their appeal. Persian cats tend to be less active than are other kinds of cat, and some may appear to be unusually tolerant of other cats and unfamiliar people, although this may simply be because it takes more to persuade them to take flight. Siamese and other oriental cats can be more dog-like than the average cat, actively seeking contact with people, and many are very vocal, seemingly trying to "talk" to their owners. Bengal cats—a hybrid with a different species, the Asian leopard cat, and originally bred for their striking "rosette" markings, certainly not their temperament—are notoriously hyperactive, and some can become aggressive toward other cats. Although most cat breeds are defined—and judged—by their appearance, one group of breeds, the Ragdolls, have been bred specifically for their extremely placid behavior. Such breeding has resulted in a relaxed and gentle cat that is extremely tolerant of handling, including being picked up: the naturally relaxed disposition of such cats should provide a good foundation for successful training.

The characteristics of the breed will need to be taken into account when training any kind of pedigree cat. Persians and related flat-faced breeds (Exotics) are reputedly somewhat difficult to train— their overly "laid back" nature can be challenging, not only in finding rewards motivating enough to engage the cat's attention but also because their flat faces can make eating small rewards from the hand or floor awkward and time consuming. Siamese and other oriental breeds can be particularly attentive to their trainer, and the challenges involved in training such breeds are more likely to stem from preventing the cat from getting bored, frustrated or overexcited. If your cat is a pedigree, you may need to research the breed to find out what behavioral quirks and personality traits may be common for that breed and plan your training accordingly.[1]

Even nowadays, most pet cats have no pedigree but nevertheless are recognizably different in personality. Some cats are loners, preferring their own company most of the time, seeming merely to tolerate contact with people or other cats. Others are very much "people" cats, maintaining a close relationship with one or more persons in their household. Less common are cats' cats, those that actively seek the company of one or more specific cats, but there are a few that are extremely outgoing, attempting to make friends with everyone— even every cat—that it meets. Most, of course, fall somewhere between these extremes.

FROM A TRAINING PERSPECTIVE, PERHAPS THE MOST FUNDAMENTAL aspect of a cat's personality is how bold or timid it is, irrespective of the situation it finds itself in. Some cats are nervous of specific events—fireworks, for example, or unfamiliar people coming to the house—but seem generally relaxed and inquisitive the rest of the time. Although this is also a part of their personality, it is also the result of particular experiences that the cat has had—and therefore a product of learning. Boldness, and its opposite, timidity, can be recognized as more general attitudes toward life: bold cats are just much more inclined to get involved in situations they haven't encountered before, finding objects new to the household interesting and exciting to explore, while timid cats hang back, finding unfamiliar objects daunting and potentially threatening—they may even run away as soon as they start to feel uncomfortable. This difference will affect which rewards to use in training. For a bold-tempered cat, the opportunity to play with new toys should successfully reinforce certain behaviors, motivating the cat to perform them again, whereas the same new toys may have the opposite effect in a timid cat—for timid cats, rewards should be familiar and therefore "safe." Of course, these are extremes—most cats fit somewhere in between.

Whether a cat turns out to be generally bold or timid depends partly on how bold or timid its parents were: in other words, there is a genetic component involved. It makes sense that some cats should be fundamentally more timid than others, for no kitten can predict what kind of world it will grow up in. If through bad luck it finds itself in a hazardous environment, a little caution may ensure it survives where a bolder cat would perish: however, if it grows up in a low-risk environment, it will probably lose out to more confident cats that are able to take possession of the available food and other resources more quickly. Thus, over the millennia of the domestic cat's evolution, natural selection has not weeded out either the genes for boldness or the genes for timidity, because both have been useful in different places and at different times in the past.

Just because boldness and timidity are influenced by genes does not mean that they are fixed for the whole life of the cat. Kittens that are bold when they leave their mothers tend to remain so for the next year or so, likewise kittens that are more timid than the average, but these differences seem to fade during the cat's second year of life. It is not known whether this is due to the effect of the genes diminishing or because cats change the way they react to the world based on the situations they have encountered, but both probably play a part. This means that there is plenty of scope to use training exercises to make a timid cat more confident or, potentially, to make an overbold cat a little more circumspect, although the latter is less likely to be a problem—to cat and to owner—than the former.[2]

How bold a cat is doesn't just affect how it reacts when confronted with a novel situation, it also affects how and what it learns. This can start at a very early age. The most timid kitten in the litter is often the one that gets the least handling, because it is easily overlooked in favor of the bolder kittens that are always the first to ask to be picked up and stroked. Although being somewhat timid may have been a valuable survival strategy in the wild, it is generally less useful in the much safer world that most pet kittens are born into. Gentle handling can overcome the immediate effects of timidity, so ideally the mother's owners will have made sure that every kitten in the litter has received its fair share of handling. If this has not happened, some remedial training may be required after the kitten is homed, to bring it out of its shell (so to speak). Nervous cats can be trained to relax while being stroked, and this may eventually result in stroking becoming rewarding in its own right, broadening the possibilities for further training.

Plenty of gentle handling during the first few months of life can turn a naturally bold kitten into a highly sociable one. Such cats are likely to find social interaction such as gentle praise and stroking rewarding, whereas cats that are very timid of people may perceive such interaction as punishing rather than reinforcing. Likewise, confident cats are likely to learn quickly and therefore need fewer training sessions, while shy or timid cats may need to work at a slower pace, with the training goal broken down into smaller and more achievable steps.

As well as its personality, a cat's readiness to learn is affected by its mood at the time. It is therefore vitally important to be able to recognize what moods are normal for your cat and also what mood your cat is in at a particular time. For example, some cats are generally difficult to motivate, almost as if they find training boring. If your cat seems lazy and uninterested, progress during training will be near impossible and may reduce both your own and your cat's enthusiasm for training in the future. At the other end of the spectrum are those cats that are particularly excitable. Training a cat that is overexcited can be as difficult as training one that is disinterested, because the cat spends the majority of the time too focused on trying to get the reward without considering what is the correct behavior to obtain it.

Luckily for us as trainers, and contrary to popular belief, cats give away their moods through their behavior and body language: this enables us to decipher what they may be feeling and to adjust our training to maximize their learning. Cats that are underengaged with a training task often show interest in anything but the trainer and the rewards on offer: they may slowly turn their heads away from you or move away completely. They often appear to want to rest, perhaps flopping slowly onto one side, or may show their disinterest by yawning or falling asleep. They may groom themselves in a rhythmic and systematic manner—note that this is quite different from the short bursts of grooming of just one part of the body that are commonly seen during momentary frustration. However, for those underengaged cats that usually enjoy touch, their relaxed demeanor will mean they are receptive to being stroked.

Luckily, there are a number of things we can do to enhance engagement in training. First, before trying to continue with any training, reassess the environment for distractions and, where possible, remove or minimize such distractions. Then consider increasing the value of your rewards—for example, using tastier food such as freshly cooked meat or fish or a more engaging toy such as a wand with real feathers or fur. Lengthen the gap between your cat's last food or play session or social interaction (whichever you are using as the reward) before commencing the next training session, thus creating greater motivation. Furthermore, reducing the difficulty of each step in the progression toward your training goal will allow you to increase the rate of delivery of rewards—and more frequent rewards will help keep your cat motivated. You can also add more variety to the rewards on offer by changing the type, size and delivery of reward often—for example, for one correct behavior reward with a piece of ham, for the next, the reward can be play with a feather, the next can be three high-quality cat treats scattered on the floor for the cat to scavenge, and so on. Scattering also encourages movement that can motivate a cat to stay focused.

If your cat really isn't interested in learning one particular behavior, try training a different behavior—perhaps a behavior that is more exciting and fun, an action that your cat already performs spontaneously, or something that involves movement and physical contact with you—for example, touching your hand with his nose or paw. Once you have gained his attention, you may find you can then move back to your desired training task. Try to always make yourself the most interesting thing in the room: move around, be animated and use your voice to engage the cat. Finally, avoid too much repetition of the same actions and keep training sessions short—leaving your cat wanting more will always help to maintain engagement over the long term.

At the other end of the spectrum are those cats that easily become overaroused: this can stem from a variety of emotions but commonly occurs in cats that are frustrated because they want something they

cannot yet get. During training, we generally want to avoid over-arousal, as cats in this mood have often lost much of their self-control and as a consequence will struggle to learn the task you are trying to teach. For example, cats in excitable moods (perhaps because they know you have a food reward) can find the presence of food over-enticing. This can lead to further excitement that, in turn, makes it difficult for them to focus on the behavior they have to perform to obtain the reward. Instead, they may spend a great deal of time me-owing and trying to obtain the food reward. Such excitement can quickly change to frustration in the cat, and as a result, many cats lash out with their claws in attempts to gain the food.

Thus it is really important to be able to recognize the signs that a cat is starting to become overaroused and thus losing focus on the training task, if only so that your training remains safe for both you and your cat. Furthermore, recognizing the signs that your cat's arousal levels are increasing will help you avoid the situation where your cat starts behaving in an undesirable way, such as "mugging" (e.g., biting or scratching) you for the food reward. If such undesir-able behavior is prevented and the cat is simultaneously taught a more appropriate alternative, the undesired behavior may eventu-ally disappear from the cat's normal repertoire, even in the face of increasing arousal.

Overarousal can show itself in many ways. Some cats will in-vade your personal space—for example, by jumping on your lap uninvited—and may scratch or bite your hands in attempts to get the reward. Other cats may meow excessively, vocalizing much more than they do when relaxed; others may display signs of agita-tion, such as dashing around, circling and quickly switching from one type of behavior to another. They may even appear to be play-ing with an imaginary toy, pouncing on thin air. Generally, over-aroused cats appear to lack concentration, which can be interpreted (unkindly) as "selective deafness." Telltale signs that a cat is becom-ing aroused include enlarged pupils and eyes that dart from side to side, the tail beginning to swish and the skin to ripple and twitch,

Sheldon is displaying "mugging behavior," biting at the lid
of the training toolbox to get at the treats inside—an example
of overarousal associated with training.

muscles becoming increasingly tense and breathing quickening. In frustrated cats, you may also witness the ears flattening slightly as they swivel backward, and the tail may begin to twitch.

Because cats differ in how they show that they're becoming over-aroused, it is a good idea to learn to recognize your cat's individual signs of arousal and the usual order in which they appear. If during training your cat's arousal levels appear at what is typically the higher end for him or appear to be continuing to increase despite attempts to reduce them, take a break. Then keep future training sessions short, finishing before arousal can escalate. Give your cat appropriate outlets for his energy at times outside of training: providing ample opportunities for play and exercise will help reduce the chances of arousal reappearing during training.

To reduce arousal levels further, you can try decreasing the value of your food rewards. For example, switch from freshly cooked meat or fish to dry cat biscuits, and avoid high-intensity rewards such as exciting play with toys. Training in as calm and quiet an environment as possible and keeping your movements slow and deliberate will also help to prevent further escalation. By teaching your cat to perform a relaxed settle (Key Skill No. 6 in Chapter 3) prior to commencing, you can be sure he starts training with the right level of engagement, thereby reducing the chances of overarousal.

Another helpful hint is to deliver your rewards in a manner that teaches your cat self-control. For example, you could drop your food rewards into a small container such as an old yogurt pot strategically placed at arm's length from you. For some training tasks it is possible to stand up, rather than sitting or crouching down, thereby making mugging behavior more difficult. As well as protecting your hands, delivering food away from your body teaches your cat that he has to move away from you to obtain the reward and also that overly exuberant behavior will make it harder for him to obtain the food. For example, he will be able to obtain the food from the pot only by carefully inserting a single paw. Commercially available automated treat dispensers are also an option. Food held between tongs or in a squeezy tube is better than hand feeding, again to protect hands from any mugging behavior. When food and toy rewards are not in use, keep them out of sight and enclosed (to stop enticing smells from escaping and increasing excitement) until ready to dispense. Although we can teach cats over time to manage their arousal levels around food, they do find it harder than dogs to switch from an aroused to a calm state.

Finally, cats also vary in what they find most rewarding. There are few cats who don't respond to tasty morsels of food, but timid cats may find playing with a "fishing rod" toy more stressful than pleasurable, and cats that are either generally timid or have not been well socialized to people may feel somewhat ambivalent about being stroked. However, this may change as the cat matures and may also be directly affected by training: for example, we will show

Ideal engagement with training

Behavioral Signs

- Your cat is more interested in what you are doing and what rewards you have than in anything else around him.
- Calm focus is on you—cats generally do not use fixed eye contact when they are feeling relaxed, so your cat will not be looking at you at all times, but he should glance at you and at the place where the rewards are.
- Your cat is close to you but not mugging you for the reward.
- He is likely to offer several different kinds of behavior spontaneously.
- Progress is being made in training with individual training goals being met.
- For cats socially oriented toward people, affectionate behavior is directed at you, such as facial rubbing.

Cosmos shows interest in the treat in my hand and is not distracted, despite being outdoors.

Sheldon is calmly focused on me—the ideal amount of engagement with training.

(continues)

Ideal engagement with training *(continued)*

What Can You Do to Ensure
Your Cat Remains Engaged in Future Training?

- Note down what circumstances have led to successful training (e.g., rewards used, size of training steps, duration of training session, time from previous meal/game to training).
- Teach your cat a command for engagement with you: because involvement with a trainer is a learned behavior, particularly for cats, who are not as socially oriented to humans as are dogs, it may be a good idea to teach your cat that engagement with you is what you would like during any training session. This should not be attempted until you can reliably get your cat to engage in training. It can be done by first saying his name (to get your cat's attention) and then using a cue word or phrase—this can be anything you like, for example, "let's work." Then, reward your cat for directing his attention toward you. The cue doesn't have to be a word: for example, it could be getting the training toolbox out and, when the cat is paying attention, opening it and providing the cat with a reward from it. This works in much the same way as the cue used to indicate that a training session is going to end—for example, a verbal "finished" or putting away the training toolbox. As your cat comes to enjoy his training with you more and more, not only will he become increasingly absorbed in it, but also his bond toward you will strengthen.
- As engagement increases, you can increase the difficulty of the tasks you wish to train. You can also generalize simpler training tasks to more distracting environments—for example, once you have taught your cat to come when called in the quiet of your home, you can then generalize this to the more distracting environment of your garden.

---

you that cats can be trained to relax while being stroked, and this may eventually result in stroking becoming rewarding in its own right, broadening the possibilities for further training.

Because cats are learning all the time (not just during training sessions), every cat will carry with him the memories of previous experiences, both good and bad, and these can have a powerful effect on how he reacts to specific situations or training techniques—in addition to the more general effects of his personality. Therefore, if you know of any relevant previous negative experiences, make a mental note not to inadvertently put your cat in a situation that might remind him of these. Teaching your cat that a previously negative experience can actually be perceived as positive often takes a long time, and it needs to be carried out at a slower pace than training where no prior experience of the task or context has occurred. If that negative experience involved a piece of equipment such as a cat carrier or a cat flap, consider changing the equipment to another model before commencing any training, so as to reduce the negative association. If purchasing new equipment is not possible, wash the item thoroughly with enzymatic cleaner (or a homemade solution of 10 percent laundry detergent with enzymes, rinsed and followed by a rub of alcohol on cotton wool onto the equipment, which must be allowed to evaporate before coming back in contact with your cat) to remove any scent traces that your cat (or any other cat) may have deposited on the item when in a negative emotional state. Such scents are likely to have been deposited from glands in between the cat's toes and on his face, and will be undetectable by the human nose, but if they are not removed they will remind the cat of its previous negative experiences with the item. For cats that have been adopted from animal shelters, past experiences are likely to be unknown, so proceed with caution.[3]

THE CASE HISTORIES THAT FOLLOW ILLUSTRATE HOW DIFFERENCES among individual cats can determine the best approach to training a particular task. The underlying learning was essentially the same for both cats, but the successful outcome relied on careful consideration of each cat's personality, age, health, previous experiences of being restricted to a small space and general likes and dislikes.

Marmaduke is a confident, inquisitive kitten whose favorite game is to play in the cardboard boxes that his thoughtful owners have filled with scrunched up newspaper, balls and feathers for him to pounce on. He loves his food and readily takes commercial cat treats from his owners' hands. They have been keen to get him acquainted with the cat carrier, as they have a holiday cottage that they visit regularly and they would like to start taking Marmaduke with them. He has been in the cat carrier only twice since they have had him, the first time when he was brought home from the animal shelter, and then a trip to the vet for his second vaccination. His owners reported that during both of these events he meowed and pawed through the carrier during the journeys, but they judged his meows to have been attempts to gain their attention rather than because he was scared. Thus, his love of boxes, food and toys, combined with his confidence around people all suggest that he should easily be able to learn that the cat carrier can actually be a fun place to be in. First, I advised his owners to put one of Marmaduke's most favored blankets inside his carrier so that it smelled familiar as well as appearing comfortable and enticing. I then instructed them how to place food treats on the floor in a trail leading into the cat carrier just before Marmaduke was due a meal. Marmaduke's curiosity and hunger were enough for him to follow the food trail, eating as he went, ending with him walking straight into the cat carrier without a backward glance. Once he was in there, the owners were asked to drop more food into the carrier through the slats in the plastic sides. Marmaduke soon learned that being in the cat carrier led to special treats. Furthermore, I showed his owners how they could play with Marmaduke while he was in the carrier, by sliding long feathers through the slats and wiggling them for him to bat and pounce on, thus providing a second reward for him for being in the carrier. A week later, Marmaduke's owners reported that he loves his cat carrier and is often heard playing with his toys in there of his own accord. Their next step will be to teach Marmaduke to remain calm when the cat carrier door closes and that even when it is moved and goes to new places, it remains a safe and fun place to be in.

Jade is a large but nervous elderly arthritic cat who has on many occasions been pushed into a cat carrier far too small for her. Being thrust into such a confined space, coupled with the pain caused by the strain on her arthritic joints, has caused Jade discomfort and distress on numerous occasions. As a result, she has strong negative feelings about the carrier and resists entering with tooth and claw. Therefore, the first task was to reverse this well-established negative association. A new, larger cat carrier was purchased to give Jade ample room to turn around comfortably. Several weeks were spent teaching Jade that this was a safe and comfortable place to be in. In the initial stages, the roof of the cat carrier was completely removed and the base was placed in the same room as her bed. Microwave-able heat pads were placed inside, under some of her favorite bedding, to make the inside warm, cozy and desirable. Highly palatable treats such as prawns and tuna were fed to Jade in her original bed, which was gradually moved closer and closer to the carrier base so that she herself had to move closer to it to get to them. Many repetitions of this finally led to a breakthrough where Jade stepped into the carrier base. This was immediately rewarded with gentle praise and hand-feeding of prawns in the carrier. From that day, she began to use it to sleep in, alternating with her original bed. Over a period of six weeks, the roof was gradually added, as was the door (which was always left open). Once comfortable with going into the carrier of her own accord for a snooze with both roof and door attached (door still left open), Jade's owners were instructed to pick times when Jade was relaxed in the carrier and then to shut the door for just a few seconds (no longer, to prevent Jade from feeling trapped), at the same time offering her a food reward through the slats of the carrier. This taught Jade that the shutting of the door results in the positive consequence of a food reward. Her owners now have Jade at a stage where she is content in the carrier with the door closed, and they are able to lift it up and take it to the car without upsetting her. The next stage will be for Jade to learn that traveling in the car in the carrier is nothing to be feared.

Marmaduke and Jade both successfully learned positive things about the cat carrier through the process of operant conditioning— that is, they found that being in or near the carrier led to positive consequences. However, there were several differences between the cats that needed to be accounted for in the training, so as to ensure that both cats reached the same goal:

- Personality—Marmaduke was bold while Jade was timid.
- Age—Marmaduke was a young kitten while Jade was an elderly cat.
- Previous experiences—Marmaduke had relatively little experience with the cat carrier prior to his training and such experiences were relatively neutral, while Jade had many experiences of the cat carrier over several years that had been generally unpleasant.
- Health—Marmaduke was in fine health while Jade suffered arthritis and therefore painful joints.
- General likes and dislikes—Marmaduke loved to play and ate anything offered to him, while Jade, as an older cat, did not play much and was fussy when it came to food.

Characterizing each cat before any training began allowed the training to be customized for each individual. For example, both previous experience and health influenced the type of cat carrier used for training each cat. Because of Jade's previous negative experiences with her cat carrier and her current arthritic condition, it was important to purchase a new and larger carrier before training commenced. Using a new carrier not only meant that Jade would be more physically comfortable once inside, with ample room to turn around, but also that she had no previous associations with this type of carrier and thus it would be easier to teach her that this carrier was comfortable. It would have been much more difficult to teach her to enjoy being in the old carrier that she so passionately disliked. For Marmaduke, however, his limited experience with his carrier, which therefore appeared relatively neutral from the outset, meant no new carrier was needed prior to training.

Age and previous experience influenced the amount of time needed to train the task. Although both cats needed short, frequent training sessions because they tired easily (as do kittens and elderly cats alike), Jade's previous negative experiences with the cat carrier meant she needed a much larger number of sessions spread over a longer time period to get to the same stage of comfort with the carrier as did Marmaduke. Indeed, her aging brain may have also contributed to the greater number of training sessions needed, as older animals often do not learn as quickly as young animals.

Finally, personality, age, health and general likes and dislikes all contributed to the decision on what training rewards to use for each cat. Marmaduke was a confident cat who enjoyed all food and play and thus a combination of food and toys were used, including the commercial food treats that he found rewarding. Jade, on the contrary, was of a more timid personality: although she had been playful when young, her current age and arthritic condition meant play was no longer high on her priority list. She had always been a fussy cat and thus fresh, protein-rich foods such as prawns and tuna were used, being the only rewards that would motivate her to go near the carrier. In addition, as Jade was an elderly cat, making the carrier a safe, secure and warm sleeping place (through the use of the familiar bedding and heat source) was most appealing to her. For Marmaduke, however, creating a place for activity and play was what made the carrier appealing.

BEFORE YOU START ANY TRAINING, YOU SHOULD ALSO CONSIDER what *your* priorities are. You, the owner, are the one who should know what your cat's needs are now and will be in the future. People's homes, lifestyles and expectations of their cats also vary, and these in turn are likely to influence what tasks are priorities for training. Thus, as well as identifying your cat's idiosyncrasies and resulting personal needs, it is important to consider what your cat's lifestyle currently entails and whether this will change in the future. You can then make sure your training helps to prepare your cat well in advance of any changes you may be planning, as well as to cope with any unexpected events he encounters. Cats, lacking our foresight, are likely to be completely unaware of these until they happen. For example, you may be acutely aware that you've arranged for builders to come in and work on part of the house and will need to close off certain rooms. The first the cat will know about this will be when you start to move furniture around the evening before the builders arrive: a timid cat will find this unsettling enough—let alone the noise, smell, dust and disruption that begins just a few hours later. The extent to which you need to prepare your cat for this will depend on his personality but also on the physical layout of your house—for example, is his cat flap well away from the disrupted area, allowing him to come and go as before, or do you need to teach him a new way in and out of the house? Other changes will inevitably require some preparation, whatever your cat's personality, such as the arrival in the house of a new cat or a new baby.

There is a lot to consider before training even begins. You will need to identify and understand your cat's needs as an individual and try to predict how they might influence your cat's ability to learn and what you need to teach, not just for now but for the future. Fortunately, by putting in this preparation you will be doubly rewarded—first, by setting up your training experience for success, and second, by also getting to know and understand your cat even better than you did before.

CHAPTER 3

Our Training Philosophy
Mastering the key skills

YOU ARE LIKELY EAGER TO BEGIN SOME TRAINING, NOW THAT you understand that you need to tailor your approach to your cat's personality and individual quirks. You probably already have a mental wish list of what you would like your cat to be able to do—whether that is to accept a worming tablet without a battle, learn alternatives to hunting, or be a little more accommodating to your visitors. But you may feel uncertain exactly where to start or how to go about it.

The basic philosophy behind all cat training is to utilize the theory of operant conditioning to reward desired behavior and ignore unwanted behavior or redirect it toward a more appropriate target. For example, you may want to direct scratching away from the sofa and toward the purpose-built scratch post that you've bought.

We will never advocate force or punishment. When a cat spontaneously offers the desired behavior, a reward should follow immediately. When he doesn't, we can "suggest" the right behavior using a number of standardized training techniques, common to all animal training, from dogs to marine mammals—they just need a little adaptation to accommodate the cat's biology and unique way of thinking. Using this philosophy, the number of things we can train our cats to do is potentially vast: so long as they have the physical ability to perform the behavior, there is no reason why it cannot be taught.

Of course, most owners will wish to train only a handful of useful tasks, but it's comforting to know that the potential is there.

Regardless of the task to be trained, there is a small number of key training skills that form the foundations to all cat training— nine in fact, just like the number of lives a cat is supposed to have. Not every task will require every key skill (although some are an essential part of all training), but, just as staple ingredients turn up in different combinations in many recipes, different combinations of these foundational training skills will be used to train different tasks. Having both a solid understanding of the science underpinning these key skills as well as the ability to successfully execute them will allow you to be flexible in your approach to training. By utilizing different key skills for different tasks and having the confidence and ability to try a different approach if one way doesn't appear to be working, you and your cat will succeed in reaching your desired training goals. Thus before diving into training the specific task you really want your cat to learn, taking the time to learn and rehearse the Key Skills using the following suggested practical tasks will set you and your cat up for successful training.

THE NINE KEY SKILLS THAT
FORM THE FOUNDATION FOR TRAINING CATS

Key Skill No. 1: Reward spontaneous observation and exploration

In everyday life, cats encounter many new things, whether they be people such as a new partner, housemate or baby, animals such as a new cat or dog, or objects such as nail clippers, a collar, a piece of furniture or a household appliance. In training terms, we call the target object, person or animal the *stimulus*.

There's a right way and a wrong way to introduce a cat to something new. First, the wrong way. It is not uncommon for owners to lift their cats and physically place them next to or even on top of novel items in an attempt to acclimatize the cat to the chosen stimulus. For example, cats are often lifted up by visitors or placed on their laps, upon which the cat tends to hastily free himself of the unwelcome attention and make a speedy exit. Another common example involves owners physically placing their cat in the new litter box or cat carrier before the cat has even had a chance to suss it out. Being in such close proximity to a new stimulus, without first observing it from afar and deciding whether it is safe, will cause most cats, particularly those of a more timid disposition, to flee, avoiding both the stimulus in question and the person who picked them up.

Cats are the ultimate control freaks—part and parcel of having evolved from a solitary ancestor for whom risk of injury is very costly. In fact, numerous scientific studies have shown that cats cope much better and show less stress—both through their behavior and their physiology—when they feel in control of a situation. Training will therefore always go best when the cat perceives it is in control of the situation and can exit from it at any time. This can be put into practice by always giving the cat the chance to observe the stimulus from a safe distance (several cat body lengths away), thereby allowing the cat to determine whether it needs to (a) flee as it perceives its safety is at risk, (b) ignore the item as it is of no threat nor interest to it or (c) approach, often with caution, to explore further as the item may hold some form of reward or the cat may need more information to make a decision about the stimulus. Whether a cat decides on outcome a, b or c very much depends on how safe it feels in its environment—for example, the outcome will likely differ for a cat in the safe confines of its home and the same cat in the daunting location of the veterinary hospital. Furthermore, its unique perception of the stimulus will be influenced by a range of other factors, such as the personality of the cat and its previous experiences with similar stimuli. For example, a timid cat with previous negative experience of dogs is unlikely to approach a new puppy.[1]

Now, the right way. When you notice your cat calmly observing, approaching and exploring something new, you can reward each action individually so that the cat associates its positive behavior toward the stimulus with a pleasurable outcome. You may be tempted to move the stimulus closer to the cat, perhaps by edging it a few centimeters at a time, but by moving the stimulus into the cat's own personal space, the cat's sense of control over the situation will diminish; therefore, this should always be avoided where possible. When a cat does respond by fleeing (even when the stimulus was presented from a distance), the next introduction needs to be from an even greater distance away, and, if possible, the importance of the stimulus, so far as the cat is concerned (its *salience*), should be reduced in advance: this can be achieved using Key Skill No. 2: systematic desensitization and counterconditioning.

If the cat's response is to take a quick look and then ignore the stimulus (something you wish the cat to interact with, for example, your cat carrier or the cat flap), you can use Key Skill No. 3, luring, which will teach the cat that interacting with the stimulus leads to rewards. If the cat spontaneously, calmly and confidently approaches the stimulus, you are on a roll—training has begun well and you should be ready to quietly deliver rewards to encourage further interaction.

Practical task

Turn an item you currently have in your home into something novel— for example, turn a chair upside down while your cat is out of the room and see how your cat reacts when he enters the room. Observe what your cat does: Does he hold back and observe before approaching and exploring through sniffing, or does he turn on his heels out of the room? Reward any positive behavior toward the upturned chair, such as calm observation, approach and exploration. If he does flee, be sure to turn the chair the right way round and allow him to return to the room in his own time. If you have a cat that showed he was scared in this way, Key Skill No. 2 will be vitally important to you.

Key Skill No. 2: Gently, gently, one sense at a time—systematic desensitization and counterconditioning

Some situations are perceived as threatening by all cats, while others are less easy for the owner to predict in advance. For example, almost all cats are naturally very cautious of unknown dogs or cats—encounters with either can be costly to the cat in terms of risk of injury or as a threat to resources such as food or territory. How other novel encounters are perceived will vary from one cat to another: for example, the introduction of a new baby toy that lights up and plays tunes may send some cats scarpering, while others may choose to observe cautiously from a distance, although very few will be confident enough to approach immediately. There will also be things that the cat finds neither enticing nor threatening. For example, most will not even bat an eyelid at a new picture hanging on the wall or a new cushion on the sofa, even though they may be the first thing that a person coming into the room comments on.

Situations that you are confident your cat will not feel frightened or worried about can be introduced to your cat in their entirety, albeit initially from a distance your cat perceives as safe (see Key Skill No. 1). If you think that your cat is likely to show a negative response to a situation—either because by its very nature it is something that any cat would perceive as threatening or because you know that your cat has previously had a similar experience and not responded positively—or you are unsure about how your cat will respond, it is important that you reduce the salience of the situation for your cat to a level where he shows no fear; otherwise, he is unlikely to learn to accept it. For example, if your cat is scared of the washing machine, you can reduce its salience by initially switching it on only when your cat is in another room. Repeated exposures at this low level allow your cat to habituate to the stimulus. You can then gradually increase his exposure to the stimulus while still keeping it low enough not to trigger any fear—this is known as systematic desensitization. For example, you could increase the exposure to the washing machine gradually by having it switched on its quietest and quickest cycle. Furthermore, if there is a window in the washing machine where your cat can see the washing moving around, you can cover this: some cats feel threatened on seeing their reflection in the window on the washing machine. Introducing a situation in this gradual way will allow your cat to learn in a gentle and kind manner that it is nothing (or no longer something) to fear and will also prevent your cat from becoming overwhelmed.[2]

When you are planning how to gradually introduce your cat to a particular situation, remember that his senses are both more acute than yours, and also different from them. Our primary sense is vision, and systematic desensitization programs for people usually focus around the sight of whatever it is they are scared of. For example, someone receiving systematic desensitization therapy for a spider phobia may be asked to imagine a very small spider very far away, thereby reducing the salience of the spider. How the spider sounds or smells are pretty much irrelevant as they generally are not bound up in the phobia. However, cats use their acute sense of smell all the time, both to communicate with other cats and to find their way around, so the smell of the stimulus, particularly if it is alive (for example, another cat, dog or human) is going to be particularly important. However, sound, sight and touch should not be ignored and will also need careful consideration. Often it may be more practical—and easier on the cat—to introduce him to different sensory aspects of the situation independently. For example, when introducing a new cat, the best practice is to initially introduce low levels of the scent of the new cat (see Key Skill No. 7), before hearing sounds made by the cat (for example, by hearing the cat meow through a closed doorway). Then, once he is familiar and comfortable with both the smell and sound of the new cat, we would let him see the new cat, reducing the impact by allowing viewing only through a glass door or from some distance away.

Although desensitization will stop your cat from feeling fearful, it will not teach him to view the situation positively. Luckily, we can also use counterconditioning alongside desensitization as a powerful way of changing the cat's perception from negative (or ambivalent) to positive. Counterconditioning involves associating what was a fear-eliciting situation with something positive. When combined with desensitization, the stimuli are presented at a level below that which elicits fear, thereby giving the animal the best chance of learning that the reward on offer is associated with that stimulus rather than anything else. Even a small amount of fear will block any pleasant associations from building up.

I used to have a cat called Harry who was scared of the vacuum cleaner. For a cat that has had no opportunity in kittenhood to learn that vacuum cleaners are nothing to fear, there are a number of sensory properties that can become linked to such fear developing in adulthood. First, vacuum cleaners can be seen (a box on wheels with a long bendable tube and brush attachment that is moved in many directions both on the floor, and at times, in the general direction of Harry); heard (a loud suction noise that starts and stops unexpectedly and an equally unexpected whirring sound when the cord retracts back into the body of the vacuum cleaner); and smelled (although almost imperceptible to the human nose, it is likely there are odors associated with the heated plastic and the dust particles being expelled through the air filter). Harry had never touched a vacuum cleaner, as he had never been brave enough to get that close. Therefore, considering Harry as a cat and an individual, it seemed likely that it was the sound (loud and unpredictable) and the movement (unpredictable and at times toward him) that were causing his fear.[3]

Harry's systematic desensitization and counterconditioning plan therefore involved initially just hearing the vacuum cleaner (on its lowest suction setting, which also reduced the volume) from another room, with the door in between closed to muffle the sound as much as possible. At this intensity Harry appeared to show no fear, although he was a little more alert than normal. Using tasty treats to countercondition, I was gradually able to increase the volume (and suction) and then to open the door to the room containing the vacuum cleaner. At other times, independent of the sound, I allowed Harry to view a stationary vacuum cleaner, first in a different room

from him—for example, while he was in the hallway and the vacuum cleaner was in the living room with door opened between the
two—and then in the same room. I then progressed to moving the
vacuum cleaner (but never toward him) while it was switched off.
Naturally, all of these exposures were paired with rewards such as
treats and praise. Sometimes, while halfway through such training, I
would stop and give Harry his meal, or stroke and scratch round his
face—I knew he liked these very much.

Over several sessions, I was able to bring together the sight and
sound of the vacuum cleaner while still keeping Harry out of his
"fear zone." Initially, the vacuum cleaner was on only at its lowest
setting while I moved it slowly. This was gradually built up over repeated exposures to normal suction setting and being moved in all
directions. Not long after such training I could happily vacuum in
the same room as Harry, who would generally stay resting on the
sofa or in his bed. I also gave Harry a warning that I was about to
start vacuuming by switching the cleaner on and off for a second
before switching it on in earnest. If he appeared disturbed by the
noise, he sometimes sauntered out of the room, but in a way that
showed he was off to find something more exciting to do, rather
than scarpering out of fear.

Practical task

Think of something that you know your cat is scared of and write down
all of its sensory properties—how it smells, feels and sounds, as well as
how it looks. Next, highlight all those properties that you think may be
contributing to your cat's fear. Remember to consider the natural biology of cats as well as your cat's own characteristics (both the introduction and Chapter 1 will help you with this). Finally, develop a systematic
desensitization and counterconditioning plan describing ways you can
reduce the intensity of the fear-eliciting properties and then how you
plan to systematically present them to your cat in a gradual way, so that
your cat doesn't feel fearful. Now, begin to work through this plan,
keeping plenty of rewards at hand for the counterconditioning.

Key Skill No. 3: Luring

Just because a cat has learned that a situation is nothing to fear does not necessarily mean he will then want to interact with it. This is not a problem for those situations that we want our cats simply to ignore—for example, the vacuum cleaner. However, there are some stimuli that we do need our cat to be comfortable interacting with; these range from simple things such as collars, harnesses, medication, grooming brushes, cat carriers and cat flaps to the more complex, such as a person—perhaps a cat-sitter—or wholly novel environments—for example, a new house or a boarding cattery.

If your cat never approaches or interacts with new things, he will have no way of knowing whether they are relevant to him or whether interaction with them may actually bring about a reward. For example, a cat who is perfectly comfortable sitting next to his newly installed cat flap (having previously always gone outside via the back door) may not know that he needs to push his head through the flap to be able to use it. Likewise, a cat may seem perfectly relaxed when you place an unfastened collar in front of him but have no inclination to push his head through the opening of the collar being offered, as he does not yet know this behavior will be made worth his while. In such cases, you will need to provide your cat with guidance as to how to behave around these objects in order to receive reward (whether the reward is ultimately access to the outdoors, in the case of the cat flap, or a food treat in the case of the collar).

Luring is usually the best way to entice the cat into any situation we want him to learn about. A lure is, just as the name suggests, something that entices or tempts the cat—thus food is the most commonly used. A tasty morsel of food placed just in front of the cat's nose and moved slowly away will encourage the cat to follow as he attempts to obtain it. Although this provides the basis for a force-free method of guiding a cat's behavior, using food in the hand as a lure can be problematic. Many cats will be tempted to swipe at the food in attempts to gain it—a natural piece of hunting behavior—especially if they are new to luring, very food oriented or have already been allowed to swipe or grab. Once training has become a well-established part of the cat's routine, it is possible to train a cat to control such excitement around food, but while you and your cat are first learning this key skill, it is important that your hands are protected from accidental contact with claws or teeth.

Fortunately, there are two other, much safer ways to present a lure. The first is to use a piece of equipment in which the lure, usually a food reward, can be placed, and then moved in a manner that the cat will follow, while the hands are kept out of the way. These include food tongs that can grasp a small food reward such as a meaty stick treat or even a single biscuit; spoons with elongated handles, which are particularly useful for wet food such as meaty chunks (baby spoons are ideal as they generally have an angled handle that helps keep the hand well out of the way); and syringes and squeezy tubes (those with an attached spoon designed for baby weaning are particularly well suited as they minimize spillage), which can be held at one end and pressed to allow liquidized food to dispense from the other end.

You will probably find it more productive in the long run to lure your cat using pastes rather than a treat in a pair of tongs or a meaty chunk on a spoon, because pastes allow you to provide a continuous but small supply of food: in that way you should prevent your cat from feeling any need to snatch or grab the food. Although this may initially appear somewhat trivial, teaching your cat from the outset that he does not need to grab or swipe at food is a very important lesson for him to learn, one that not only protects your hands but also teaches your cat how to control his desires. You can make liquidized food by puréeing wet cat food or meat and fish pastes with a little water. If using meat or fish paste meant for human consumption, be sure to check first with your veterinarian that the ingredients are safe for feline consumption. Commercially available cat pâtés obviate the need to purée or mash, and they can also be easily loosened further with a little water. Learning self-control has many advantages for a cat living among people; many things he may desire will not always be immediately available, whether that be food, outdoor access or attention. Owners often complain of cats vocalizing excessively or scratching the furniture; both of these nuisances can result from the cat feeling frustrated and lacking self-control when he is unable to get what he wants as soon as he wants it. Thus, teaching your cat to be patient may improve more than one aspect of your relationship.

The other method of luring involves the use of a target stick, which is a thin retractable pole with a small ball on the end, often made of foam or plastic—the "target." The idea is that the ball is an easily seen object that, by nature of the cat's curiosity, he will likely want to sniff. If he doesn't initially offer this behavior, a little food can be smeared on the target. If his inspection of the target is rewarded, he will likely perform it again. Over several repetitions of rewarding the behavior of his nose coming very close to or touching the target, you can gradually start to move the ball, initially a slight distance. Your cat should follow, and if he does, you should reward this behavior. Using these principles, you can begin to move the target stick greater distances before removing and rewarding, and your cat should still follow. Always move the target stick out of your cat's reach while you reward the behavior—for example, place it behind your back: this will prevent your cat from touching or following the target at a time when you are not ready for him. Target sticks are available commercially, but it is also easy to make one, simply by sticking a ping pong ball or something similar on to the end of an old retractable aerial or a piece of gardener's bamboo cane.

Sheldon is learning to touch a homemade luring stick with his nose.

Once Sheldon has touched it, the target stick is removed and he is rewarded with a squeeze of meaty paste from a syringe.

Practical task

Practice luring your cat with liquidized food presented in a syringe or squeezy tube. Squeeze it gently to allow a tiny amount of food to appear at the opening and then present it calmly and quietly, allowing your cat to approach, sniff and lick it. Once your cat has learned that this new contraption produces a most delicious treat, you can move it away from him by only a centimeter or so, and wait to see whether he follows. If he does, once again squeeze gently, releasing some more food. If he does not, wait until he is licking from the tube or syringe and while still pressing it to dispense more food, move it gradually so your cat is getting small amounts of food the whole time he is moving with the dispenser. Over several repetitions, you should be able to gradually increase the distance you move the dispenser before releasing some of the food. Practice this until you can get your cat to move in a circle around your body and to move from the floor up onto a chair or raised platform of some sort, dispensing food only at the end of the task. For some cats, this will occur in a single training session; for others it may take several. Just work at your cat's own speed, and always pace the session so as to end on a really positive note—for example, when the cat is still engaged and not fatigued or bored, and has just performed a really good version of the desired behavior. Don't be tempted to continue until the cat walks away, because by then the value of the reward will have diminished.

Over several more training sessions, practice these tasks using food rewards held in tongs or on an elongated spoon. With these, only one reward can be delivered at a time before you have to "reload." However, by this stage of training this should not present a problem, as your cat should be following the food lure without the need to eat as he goes. In fact, it is good practice—to prevent grabbing and increase self-control—not to give your cat the food in the tongs or the spoon as the reward when your cat reaches the lure at its final destination, but instead quickly remove this from sight and present him with a different reward. This way, your cat will learn that grabbing at the food does not result in getting it, but following it while showing self-restraint is what actually results in getting something good.

Once you have mastered this, you can move to practicing with a target stick, which has no food within it to entice your cat to follow. However, your cat has already now learned that following something that you are holding results in a reward, so he should be primed to follow the novel target stick. Start from the basics again, rewarding any investigation of the target, such as sniffing, and gradually move the stick as the cat follows to your target location, then remove it out of sight and reward. As with the food lure, work toward luring the cat around in a circle and onto a raised platform, this time using your target stick.

Key Skill No. 4: Marking a behavior

Often during training, it is not possible to deliver the reward at the appropriate moment—say if the cat is some distance from you. Furthermore, cats, like children and dogs, can switch their attention from one thing to another very quickly, and thus their behavior can change in an instant. By the time you present your cat with his reward, whether it be a food treat, toy or stroke, he may be doing something different from the behavior you wished to reward. He will most likely associate the (irrelevant) behavior he performed immediately before delivery with the reward, because they occurred most closely together in time. The secret to successful training is to reward the behavior at the very moment it occurs, thereby giving your cat the best chance to associate the reward with the desired behavior and not with something else he happened to do a few seconds later.

Consider a cat walking toward his newly fitted cat flap. When training a cat to go through a cat flap, the first behavior we might wish to reward (for example, with food) would be simply walking in the general direction of the flap—recall Key No. Skill 1, rewarding voluntary approach and exploration. However, in the couple of seconds that it takes you to bring a food treat to the cat, it is not uncommon to find that the cat, on hearing you rummaging to get a food treat out of your toolbox, has stopped in his tracks and turned away from the cat flap to see what you are doing. What we would then actually be rewarding is the turn away from the cat flap, the exact opposite of what we wanted to reward. We therefore need a way of telling our cats, "Yes, that's it, that's the behavior that will get you a reward," at the precise moment that they perform the behavior, or at least no more than a second or two afterward. In training terms, we call this *marking* a behavior. In dolphin training, this marker of correct behavior is often a blow on a whistle, in dog training it is often the click of a little clicker box. However, it can be as simple as us saying the word "good" or "yes" out loud, in a distinctive way.

This marker, or *secondary reinforcer*, as it is more formally known, becomes rewarding because it predicts that the primary reward (or primary reinforcer: food, play or stroking) is going to be delivered, and soon. The marker doesn't start out being rewarding in its own right: this has to be learned. In apparent contradiction to the way that many owners talk to their cats, cats do not have the ability to understand elaborate verbal explanations that something they have done recently was good (or bad, for that matter). Even if we could convey this information to them, mammals' brains are not built in a manner that allows them to cope with such a long delay between what they have done and the reward they receive. Thus for cats, secondary reinforcers allow us to mark the behavior as it is occurring, buying us a few seconds in which to deliver the real reward (be it food, play or gentle stroking), by which time it does not matter whether the cat has started doing something else.

Markers need to be simple, perceived instantaneously and easily recognized as such, even at a distance. A spoken word is therefore ideal, although you should select a word you do not often say to your cat at other times and perhaps deliver it in a distinctive voice that you only use while training—it is the sound that matters, not the word itself. A mechanical "clicker" can be used with cats, as with other animals, and some people prefer using this to using a word, as they can be sure the cat hears this sound only during training and never at other times. Furthermore, the sound is always precisely the same, thus adding consistency to the training. Although clickers can be really helpful during training (and there is a wealth of information out there on clicker training), for those new to training cats, it is yet another object to hold in the hand, and it can take a lot of practice and coordination to get the timing correct. For this reason we favor the hands-free "verbal marker" when teaching people to train their cats. Moreover, a sound can't get misplaced like a clicker can.[4]

Initially, the word "good" (or whichever word you choose) will have no meaning for the cat. You therefore have to teach him that when he hears this word, a tangible reward is imminent. This is done using classical conditioning: repeated pairings of saying the word, then providing the reward. Studies have shown that animals learn best when there is a very slight overlap in time between the end of presentation of the marker word and the onset of delivery of the reward. In practical terms, this is nearly impossible to do as we would be splitting seconds. Luckily, for us, animals also learn the association quickly when the reward comes immediately after the marker.

Before training any task, say your marker word (you may need to say your cat's name first to get his attention) and then deliver the reward. The word must always come before the reward, and the time delay between the marker word and delivery of reward should be very short (ideally, less than a second) during this initial training. Repeat this exercise five to ten times within a single session, depending on your cat's attention span. It is good practice to repeat this over several sessions before integrating your new marker word into your regular training. You will know when your cat has learned that this marker word predicts the arrival of a reward when you find that every time he hears the word, he behaves in a way that shows he is anticipating a reward—for example, turning his head and gazing toward you, meowing, purring or moving closer. Before doing any training where you know you are likely to use a marker, it is a good idea to perform a few repetitions of your marker and reward pairings just to check that your cat still remembers what the marker means.

Through this process of repeated pairings, the secondary reinforcer becomes rewarding in its own right, simply because it has become predictive of a "real" reward (studies show that the cat actually becomes momentarily happier whenever he hears the secondary reinforcer).[5] If you choose to use a clicker, you can physically place it into your toolbox: if you choose to use a simple word as the secondary reinforcer, it can be "virtually" placed into your toolbox—or even literally, if you find it useful to place a note in your toolbox reminding you which word(s) you have chosen. If you have several cats, it is a good idea to choose a different word for each cat, to prevent confusion if you see the opportunity to reward a spontaneous piece of behavior in one of your cats while the others are within earshot.

Practical task

Can you teach your cat to sit? Although it may not have huge influence over his well-being, this provides the perfect opportunity to practice using a marker without worrying about damaging the training of behaviors that will be more important for your cat's well-being, such as accepting medication or enjoying grooming. Hint: if opportunities to reward spontaneous incidences of sitting are few and far between, don't forget you can use Key Skill No. 3, luring, to guide your cat into the sitting position—this will allow you to get the hang of both luring and marking in the same training session—a pairing of skills that will be common for several training tasks.

Key Skill No. 5: Touch-release-reward

Some training can be completely hands free—for example, teaching the cat to come when called (unless of course you have chosen to use physical affection as a form of reward). However, there are situations in caring for a cat that will require the cat to be touched in ways he may not instinctively enjoy, such as being gently restrained for medical procedures, having different parts of his body checked during a health examination, and having his head and neck area touched while having a collar fitted. In other tasks, an object will be used to touch the cat, such as a stethoscope placed against the chest or nail clippers against the toes. Utilizing both Key Skill No. 1, reward spontaneous observation and exploration (of the finger/hand/object to be used to touch the cat), and Key Skill No. 2, systematic desensitization and counterconditioning, in combination will ensure that the finger, hand or object loses any connotations of fear and instead is perceived positively. Once these skills have been applied, it is time to start training for the specific type of touch by the finger, hand or object that is needed to perform a specific task.

Regardless of what form the touch takes, the key point to remember is that the touch occurs in its entirety (presentation and removal) before the reward is offered: touch-release-reward. For example, we might be using systematic desensitization and counterconditioning to teach a cat to have its paw lifted as part of training the cat to comfortably wear a harness: the first stage might be to teach the cat to feel comfortable having his paw touched, before we move the goal to lifting his paw off the ground. The key skill would be, touch finger to paw–withdraw finger–offer reward. The same will be true for an object: for example, we may practice applying a spot-on treatment (used to treat external parasites such as fleas) by initially just placing the unopened tip of the spot-on treatment in between the cat's shoulder blades. The same sequence applies: touch–remove spot-on container from cat–reward. Over many repetitions your cat will learn that by behaving in a calm manner and by keeping still when the touch happens, he will be rewarded.

Why do we always need to remove the touch before the reward is given? If touch and reward are introduced simultaneously, the cat may focus his attention only on the food. Although it may then seem that he does not mind being touched, it may simply be that the food was enough of a distraction to allow him to be touched (and anyway it is not practical to have to provide a constant supply of food in one hand while trying to touch the cat with the other). When the food isn't there, he will have only the touch to focus on and may realize how little he actually enjoys it, thus demonstrating that he hasn't truly learned that it can be a positive experience. Although it may be tempting to just provide the food simultaneously with the touch because you seem to achieve the same end result—the cat keeps still and allows you to perform whatever it is you need to do—you will soon find, when you try it in a more daunting environment such as the veterinary consultation room, that your cat has not actually learned to cope with touch at all. It is likely that the cat will refuse to eat in this environment and actually may resist any touch with tooth and claw. So instead, always keep your cat within his comfort zone when building in behaviors that require forms of touch and physical interaction (using Key Skill No. 2, systematic desensitization and counterconditioning), and when it's time to reward, follow Key Skill No. 5 of touch-release-reward so your cat actually learns to accept the touch.

Practical task

Can you touch your cat's paw while he remains fully comfortable? Remember to allow your cat to investigate your finger before you place it on his paw—he may sniff it or simply look at it, but reward this investigation. Allowing him this time will prevent him from startling when you reach forward to touch his paw. If your cat is particularly sensitive, you may wish to touch the ground next to his paw first and reward your cat for staying still while your finger approached close to him. Remember, the reward should always come after the finger is removed from the vicinity. Practice until you can touch your cat's paw with a single finger for two to three seconds while your cat remains calm.

Key Skill No. 6: Teaching relaxation

So much of what we require of our cats—from being in the cat carrier to having a health check to being groomed—requires the cat to stay relatively immobile. A cat is much more likely to keep still if he is relaxed. By teaching your cat to associate relaxation with a comfortable place, such as a blanket—achieved by rewarding successive approximations of relaxed behavior—you create a situation whereby your cat is relaxed before you commence any training. A relaxed cat will learn much more effectively than will a cat that is nervous or overexcited. Therefore, several of the training tasks in this book will begin at the point where the cat is relaxed on his special relaxation blanket.

The secret to success with this key skill is to reward the emotional state of relaxation, rather than simply the physical location or the posture the cat finds himself in. For example, cats can be lying down but remain very alert and vigilant, even on edge. We want to reward the cat only when he is content and relaxed. By rewarding such a relaxed state while on a specific blanket, we create a link in the cat's mind between the blanket and relaxation. Essentially, the cat gets a double reward—having the opportunity to relax, which all cats love, and receiving a tasty morsel or a gentle stroke as well. The beauty of this simple task is that the blanket is portable, and therefore when we introduce it to new places or situations—for example, in the cat carrier or boarding cattery—the cat will feel more relaxed there with it than without.

After selecting a comfortable blanket for your cat—possibly one he has previously enjoyed sleeping on—place it in front of you on the floor. Cats are naturally curious animals, and your cat is likely to come over to investigate the blanket in its new location, perhaps sniffing it or placing his paws on it to walk across it to reach you. Reward any of these behaviors. Because our ultimate goal is to have the cat relaxing on the blanket, rewards should be selected carefully so as not to excite him too much. There is a fine balance between having rewards that are motivating enough for the cat to work for them and getting him too excited while he's anticipating their delivery and thus being unable to relax. One way to avoid the latter is to switch among several different rewards—for example, one that is calming but perhaps less motivating (stroking) can be alternated with one that is highly motivating but also a little more likely to increase excitation (food).

Let's use as an example the way I taught Sheldon to relax on a blanket. To start, I chose a time when Sheldon was fairly relaxed, having already played with me earlier in the day. I placed the blanket on the floor between us. Being a sociable cat, he naturally stepped onto it to approach me. In the initial stages of training a "relax," the behavior you desire may occur only very briefly: for example, the action of stepping onto a blanket is over in a fraction of a second. Using a verbal marker such as the word "good" at the precise time the behavior occurs can ensure that you reward such instantaneous behaviors. Remember, however, that the word "good" must have already been set up as predictive of the real reward, which should also follow very shortly afterward. In Sheldon's case, a food reward was given. Initially, I dropped the food reward in front of me so Sheldon had to get off the blanket to receive the food. This did not affect his learning that going onto the blanket was what caused the reward to appear, as the marker word "good" was given only when Sheldon was actually on the blanket. By giving him the food (or primary reinforcer) off the blanket, Sheldon was being given the opportunity to go back on to the blanket several times, allowing him to gain several more rewards and consequently consolidate his learning. Initially, Sheldon spent a lot of time around me trying to get a morsel of food, but he learned very quickly that it was the act of stepping on to the blanket that actually produced the tasty food.

Sheldon is confident enough
to place all four feet on the blanket.

The next step was to build in some time on the blanket—that is, not only stepping onto it but then remaining on it. I built this up very gradually, by withholding the word "good" for an extra second or so each time Sheldon stepped on the blanket. Repeating this, I got to the point where Sheldon would stay on the blanket for several seconds waiting to hear "good," knowing he had performed the correct behavior.

Sheldon learned really quickly using this method, but if your cat does not initially voluntarily step onto the mat, you can start by luring him on with the use of toys or food. In addition, lifting up the blanket and repositioning it after giving the reward can focus your cat's attention back to the blanket, thus encourage him to reinvestigate it. On first sight of the blanket, Sheldon stepped onto it with all four paws. However, wary cats may initially place only one paw on the blanket, and for such cats, rewards should be provided for one paw on the blanket, then two, building up to all four feet—thereby breaking down the goal into smaller achievable steps. Sheldon's first training session ended with his final reward being given to him on the blanket. Sheldon was quite excited by this point because he anticipated the food treats, but because my goal at this stage was only to get him onto the blanket, not necessarily relaxed, this was not a problem. The high-value treats had focused Sheldon on the training task, and I could subsequently switch to more relaxing rewards, knowing he was engaged and on board with the task.

Now that Sheldon was comfortable standing on the blanket, the second stage was to teach him to relax on it. I continued to say "good" and reward with food treats for being on the blanket, but each time that I rewarded Sheldon, I slowly placed the food reward on the blanket just in front of Sheldon's chin. This caused him to lower his body to obtain the treat, thus shaping his body position into the preliminaries of lying down. Importantly, any signs of increased relaxation were rewarded. These can include postural changes—for example, moving from standing to sitting, to crouching, to lying with feet out to the side and lowering the head to rest on the blanket—and also any obvious signs of relaxation—such as looking away from me, slow blinking, closing of the eyes, grooming a front paw, purring and dozing. When I gave my verbal marker "good" to let Sheldon know he was about to receive a reward, I kept my voice quiet and calm to maintain the relaxation he was showing. I also interspersed food rewards with praise, in the form of gentle stroking of his head and chin scratches, to keep Sheldon in a calm mood.

Now, Sheldon is being rewarded
on the blanket for staying on it.

It can take some time for cats to make the decision to perform a behavior; thus, if your cat does not do what you desire immediately, do not give up, but sit quietly and wait. Cats can take time to process information and act upon it. For example, Sheldon took just over two minutes before he began to close his eyes while resting on the blanket. Such a time can feel like an eternity when you are waiting, but patience is key. We practiced such relaxation over several sessions until Sheldon would quite happily settle down on the blanket as soon as he saw it.

Sarah is stroking Sheldon as his
reward for lying relaxed on the blanket.

Practical task

Have a go at training your cat to relax on a specific blanket using the steps outlined above. It is a good idea to start with a blanket that your cat already uses. It may take several training sessions to teach your cat to relax fully on the blanket. Try keeping notes of each training session—things such as how far you got to toward your end goal and what worked best to aid relaxation. These tips will help you progress quickly in your training during your next session.

Key Skill No. 7: Collecting a cat's scent

Several training tasks involve introducing the cat to something new—whether that be another cat, a new piece of furniture or even a new house. Smell plays a large part in how a cat perceives its environment and the stimuli within it. As humans, our first port of call when assessing for danger is to look around, visually scanning the environment. For example, we may notice an intruder has been in our home from visual disturbances such as a picture hanging squint on the wall or a knocked-over vase. However, we are highly unlikely to detect that an intruder has been in our home based on any smell he may have left—our noses are simply not sensitive enough. For cats, it is often the other way around: their acute sense of smell and ability to detect chemical messages from their own species (through their vomeronasal organ) makes "their" world subjectively quite different from ours. Cats will routinely patrol their territory, whether that be indoor-only or both indoors and outdoors, sniffing as they go to see whether any change has occurred since their last inspection. They will stop and sniff especially intently at boundary points, such as doorways and garden gateposts. You may have seen your cat do this and then immediately rub his cheeks on an object nearby or reach up and scratch with his front claws.[6]

Facial rubbing and scratching are both believed to lead to the deposition of secretions which convey two kinds of information—first, a "signature" that is unique to that cat, and second, pheromones, which are odors that are the same in all cats. By facial rubbing and scratching, cats can communicate to others of their own species that they have been in this particular place and possibly even how long ago they were there. It is also believed that such secretions may have a useful function in helping the cat who deposited them feel secure in his own environment through the process of physically marking his own identity onto objects, thus providing a sense of environmental security. We may get such feelings from keeping our doors locked at night, but for cats, security comes from being able to detect their own scent in their environment. Thus, when a new object, person or animal enters the home, it will stand out as not containing any of the resident cat's smell and may therefore be treated as a potential threat. Some cats, generally those of a more confident personality, may happily explore the new stimulus and voluntarily facial mark it if it is not too intimidating, but this is unlikely to be the case if the stimulus is a new animal (which may resist the advance), nor may it be easy if there has been a great deal of change to the home, such as a complete redecoration of one room.

Fortunately, it is straightforward to capture a cat's scent, while he remains content and relaxed, and then transfer this chemical information to new items to the home—just as if he had marked them himself. Having the opportunity to do this prior to introducing the cat to the item gives a beneficial head start to training. There are a few ways we can collect the scent of a cat in order to present it to another animal or apply it to an object or encompass it into the smell of a new home. The first is to wear a clean, light-weight cotton glove while stroking the cat around those areas on the face that produce the chemical secretions. These are the patches in front of the ears, where the fur is slightly sparser, and under the chin and the cheeks, which start just behind the whiskers. Alternatively, or additionally, you can collect the hair from the brush you've used to groom your cat, again concentrating on the cheeks and under the chin. If, however, your cat is not keen on either of those forms of physical interaction, you can place a small piece of cloth on his bed for him to lie on and passively impregnate with his scent. The more times your method of collection (glove, brush, cloth) touches the areas that produce the secretions, the greater the concentration of scent captured. However, to begin with, keep the scent weak by stroking or brushing only a few times or leaving the cloth in the cat's bed for just one night. More concentrated samples can be used later in training: they are collected in the same manner but simply are made more concentrated through rubbing or brushing the cat for longer—this should occur over several sessions to be certain it remains a positive experience for the cat—or by leaving the cloth in the cat's bed for greater periods of time. Having collected the scent, it can either be rubbed on the stimulus if it is inanimate (such as a piece of furniture) or if animate, such as cats or dogs, placed at a distance that they can investigate in their own time.

Collecting scent by stroking Herbie with a
light cotton glove on and around the facial glands.

Practical task

Practice capturing your cat's scent using the method you think your cat will be most comfortable with. If your cat does not enjoy being groomed, keep to the method of wearing a glove while stroking your cat. However, if your cat is also not content while being stroked, simply place a piece of cloth in his bed. Pick an object in your home to rub the collected scent against and watch your cat's behavior toward this designated object over the next few days. See if you notice any changes; for example, does your cat approach the object more, sniff it or even rub against it?

Key Skill No. 8: Maintaining a taught behavior

When you first teach your cat that a particular action it performs is linked to reward, the reward should be provided for every desired response—in other words, every correct response is reinforced. By rewarding every single time the desired behavior is shown, a strong association between the behavior and the reward is created in the cat's long-term memory. In training terms, this routine is known as continuous reinforcement. Continuous reinforcement creates a behavioral response that is both highly reliable and consistent.

Once the cat reliably produces the desired action, it is important to move away from continuous reinforcement. There are several reasons for this. First, if we reward every instance of a behavior, we have no way of refining it. For example, let's say you've trained your cat to come when called: sometimes the cat will run back to you as fast as possible, at other times he may saunter back, stopping to sniff a plant or run up a tree on the way. Although some responses may be exactly what you're after, others may not, but if you rigorously apply continuous reinforcement, all responses receive the same reward. The cat will come to know that he will always be rewarded once he makes it back to you, waiting patiently at the open doorway, so why rush if there are other interesting things to do along the way? By carrying on using continuous reinforcement, we cannot create a situation where the cat only ever trots home quickly and directly. Thus, although continuous reinforcement is good for temporarily increasing the frequency of a behavior, it is not effective for maintaining or improving the frequency and quality of the behavior.

Second, behavior that has always been continuously reinforced is particularly susceptible to extinction—that is, the behavior may no longer be offered after just a few occurrences in which the reward does not follow immediately. For example, consider the situation where an owner has always, over many years and without fail, provided a treat to her cat for voluntarily entering his carrier. On one occasion, the owner has run out of treats and forgoes the reward, thinking it won't make a difference. The cat may stay in the carrier for several seconds or even minutes waiting for the expected treat. He may then come out of the carrier and approach the owner in attempts to produce the anticipated reward. He may then go back into the carrier as a second attempt to try and get his owner to produce the treat. Once he realizes it is not coming, he may well go back to doing what he was doing before the carrier was brought into the room.

Although the majority of cats will simply stop performing the action if no reward is forthcoming, there will be the occasional cat that will respond to this situation with intense frustration. Thus, a third reason to stop continuously reinforcing a behavior once it is well established is that in some cats, if the reinforcement is accidentally or intentionally not produced, it could lead to risky behavior such as the cat swiping or even biting the owner in attempts to produce the reward. This will be more common in cats that have always been continuously reinforced in many aspects of their life or those whose personalities predispose them to easily become aroused (for example, those with a low threshold for experiencing frustration).

In all cases, the chances of the cat going into the carrier next time it is brought out have gone—in technical terms, we say the behavior of entering the cat carrier has been *extinguished*. The cat has quickly learned that going in the carrier no longer produces reinforcement and therefore it is not worth his while trying this behavior again.

Additionally, it is simply not practical to provide a reward every single time your cat performs a desired action throughout his life. For example, your nervous cat might approach and greet a visitor to your home just as you are inviting them inside—with your hands full with opening the door and embracing your visitor, you miss the opportunity to reward the cat. Also imagine the situation where you are driving your cat to the vet and he changes his behavior from sitting and being alert to lying down and relaxing in his cat carrier: you are not able to deliver the reward as you have to concentrate on driving and there is no one else in the car to deliver it for you. Finally, the provision of a reward every single time the cat performs the behavior would very quickly lead to a very corpulent cat if the reward was food or a very exhausted cat if the reward was play. If the reward used was stroking, cats generally prefer physical interaction to come briefly and often, and therefore continually rewarding with stroking may quickly lead to a situation where it is no longer rewarding, because of the intensity at which it has been given.

So, once a cat is reliably producing the desired action, in order to maintain that behavior and improve its quality, you should gradually move from a continuous reinforcement schedule to one where not every offering of the behavior is rewarded. This is known as a partial or intermittent reinforcement schedule. This tends to create persistent, longer-lasting behavioral responses that are much more resilient to extinction.

Consider how addictive slot machines, or in fact any form of gambling, can be. They all rely on intermittent or partial forms of reinforcement—not every coin placed in the slot machine or bet made results in the reinforcement of a win. In fact, gamblers can never predict which coin or which bet will lead to a win—this is what makes their behavior so persistent. When a win does come, the reward that is felt is greater than the reward felt if a payout came every time, even though the monetary value of the payouts put together is actually less.

Imagine a scenario where a cat is meowing to be let outside. If the owner sometimes ignores the cat, sometimes opens the door immediately and sometimes waits for a period of time before opening it, the behavior of meowing becomes much more ingrained in the cat and as a result will occur readily, will persist and will keep occurring day after day. In this situation, the owner has unwittingly created the situation she was trying to avoid—a cat that meows incessantly. The intermittent reinforcement of the behavior will have made it more resistant to extinction. Thus, it takes a greater time period of completely ignoring the behavior before it will finally stop. Although initially often difficult for owners to do, consistently ignoring the meowing (i.e., always making sure it has no reward) is the only thing that will stop it.

However, the same phenomenon can be used to the owner's advantage when it comes to desirable behaviors we wish to teach our cats. For example, a cat intermittently reinforced for coming when called is likely to come quickly and reliably, as he will not want to miss out on the opportunity of the reward—if the reinforcement has been intermittent, he cannot know when the next one will be offered and will be keen to come to you to find out. In this situation, intermittent reinforcement helps owners get their cat back indoors or from another part of the home quickly and reliably, avoiding them having to go searching for their cat.

As well as varying whether we provide a reward or not, when we do provide a reward we can also vary how long or how many times we allow the cat to perform the action before we give the reward. For example, sometimes we may reward relaxing on the relaxation blanket after a couple of seconds, other times the reward may not come until after several minutes. For other behaviors we may vary the number of behavioral responses we require before we produce the reward— for example, we may perform multiple strokes with a grooming brush before rewarding with a food treat on one occasion, and then reward after one stroke the next. We can also vary the type and value of reward we use: reserving the highest-quality rewards for the best examples of the desired behavior will help improve the quality of the response.

These so-called schedules of reinforcement therefore have a profound effect on training: the variety of different schedules that can potentially be used is vast and precisely defined in the psychology literature. B. F. Skinner, one of the most influential psychologists of the past century, who spent much of his career studying operant conditioning, even dedicated an entire book to the topic. However, the aim of our book is not to create optimally obedient animals whose behavior is impeccable, and therefore we don't need an extensive knowledge of all the different reinforcement schedules and of precisely when to use them. Instead, we simply want to encourage behavior (by associating it with reward) that will prevent us from having to manhandle our cats for various procedures and thus promote positive emotions and relationships with us. It is enough to understand that once a behavior is being offered reliably, moving to a partial or variable schedule of reinforcement is the best way of sustaining its performance. As long as we regularly vary when, what and whether a reward is given, we will reach success in training our cats to behave the way we want them to.[7]

Practical task

For this practical task, go back to the training exercise of teaching your cat to sit (the practical task associated with Key Skill No. 4, marking a behavior) and use this to try out offering different schedules of reinforcement for the delivery of your rewards. Before starting this new exercise, familiarize your cat with the exercise of sitting for a reward. Next, select five individual training rewards—they may comprise any permutation of different types of food reward, different toys to play with, or even some form of physical contact—as long as they are all things your cat enjoys. On a piece of paper, write out the numbers one to ten and randomly assign five numbers to the words "no reward." With the remaining five numbers still to be allocated a task, randomly assign one of your five selected rewards. You should now have a list of ten consequences that you should deliver in the order written down—this is your reinforcement schedule. Use this schedule to reward your cat for the next ten sits he performs (it is still OK to lure your cat to sit at this stage if he does not voluntarily offer the behavior). Remember that if you use a marker word (as detailed in Key Skill No. 4), you should use this word only when you are delivering a reward—thus, when "no reward" is written on your schedule, you should give neither the marker word nor a reward. Observe your cat's behavior and think about how he responded with the different types of rewards and more importantly, when there was no reward. Think about what influence the reinforcement schedule had on your cat's motivation to sit—for example, was he quicker or slower to sit at any time?

Key Skill No. 9: The end is near—how to finish any training task

Once a training session is under way, it is often difficult to know when to finish. If it is not going well, you may feel compelled to keep going until your cat "gets it": you may not want to give up and feel you have failed. Likewise, if the training session is going well, you may feel enthused and motivated to proceed to the next training goal more quickly than you'd planned. However, it is really important that training sessions be kept short, each lasting only a few minutes for cats that are new to training, and for those with more experience, certainly not longer than ten minutes.[8]

For a session that is not going well at all, it is better to stop and have a break than try to battle on. Assess whether the cat is underengaged and what you can do to improve this in your next training session. Your break may need to be only a few minutes, but it will be extremely valuable in giving you time to reassess and to stop the training session from becoming associated with negative feelings—for both you and your cat.

When a session is going well, deciding when to finish can be difficult. In this situation, it is best to stop when the cat is still motivated to perform the behavior and has just shown a great example of the desired behavior. Although it is so tempting to wait for the behavior to be offered again, often it is not offered at the same quality, and ultimately you end up in a situation where you let the training session go on too long, with the likelihood that the cat will offer as good an example of the behavior as its last dwindling, because he is becoming increasingly satiated or tired.

Once we have decided we are going to stop a training session, how do we go about it? We could just simply walk away, but this may not be a clear enough signal to your cat that no more goodies will be delivered: he may follow you and may feel frustrated at the sudden lack of reward. Also, in the early stages of training a behavior, we do not want to risk the situation where your cat suddenly offers the action you're wanting to train but you are too busy packing up to notice and consequently reward. Thus, providing a clear signal that the training session is finished can be invaluable. This signal can be something as simple as saying "all done" or "finished," or making a movement such as putting your hands up or crossing your arms. Because this signal will also be followed by the behaviors of packing up any training equipment, putting away the toolbox and moving away from the cat, the cat will quickly learn (through classical conditioning) that the chance of reward is over for now. Ensuring short training sessions that end with a clear signal will help keep your cat free from fatigue, satiation and frustration throughout your training.

Practical task

For this practical task, go back to the training exercise of teaching your cat to accept having his paw touched (the practical task associated with Key Skill No. 5, touch-release-reward). Decide on the signal you are going to use to teach your cat a training session has finished. Make sure the signal you select is appropriate for your cat—for example, if your cat is quite flighty and you choose a hand signal, make sure you move your hands slowly. Touch your cat's paw three times, then give your end-of-training signal: repeat this four times (i.e., a total of twelve touches over four training sessions). Observe your cat's behavior immediately after you have given your signal—has his behavior changed by the fourth session? Do you think he has learned what the signal means? Behaviors such as getting up and walking away or looking the other way from you or engaging in another behavior are all clear indicators that your cat has learned the training session has finished.

—It's time to get started!—

CHAPTER 4

How Cats Adapt to
Living with an Alien Species (Us!)

A LTHOUGH ALL DOMESTIC KITTENS HAVE THE POTENTIAL TO become pets when they are newborn, this is not inevitable. What domestication has given them is the ability to become socialized to people under the right circumstances. Worldwide, there are probably as many cats that don't like people as cats that do. This crucial difference stems from the kind of experiences that these cats had while they were quite young kittens. Kittens that are regularly handled in a gentle way learn to trust people and enjoy their company. Kittens that either have no contact with humans or only negative experiences grow up to become fearful of people: even though they may live out their lives scavenging from us, most will never allow a human to touch them. Such cats are often referred to as "feral," to distinguish them from truly wild animals. Although feral cats are the same species as a Scottish wildcat, and the two can interbreed, their behaviors are very different. Whereas a feral cat is descended from cats that were pets, a genuine Scottish wildcat's ancestors were all wild, whether one goes back ten generations or ten thousand.

Many cats that do end up as pets are somewhat shy or timid. These cats, unlike feral cats, will have had some positive handling as kittens but not enough to develop confident social skills with people. Some may have developed affectionate relationships with their owners but dash out of the room at the first sign of visitors.

Others may tolerate a fleeting stroke from their owner but rarely seek physical contact. Then there are those cats that appear socially confident—that is, until there is a change in the social composition of the home, the most dramatic being the addition of a new baby.

Training, using a systematic approach, can result in dramatic changes in how a cat perceives people of all kinds: the cat is taught that the presence of a person, and then interaction with people, results in positive consequences, using rewards such as food treats and play. Such changes will not only be beneficial to the cat's owner, they will also improve the cat's life, preparing him to encounter the many new people he will meet over his lifetime, whether they be fleeting visitors or permanent new members of the family.

KITTENS PROBABLY START TO LEARN ABOUT HUMANS AS SOON AS their eyes and ears start functioning, which is when they are about two weeks old, and the first phase of this social learning continues until they are eight weeks old. Even a small amount of gentle handling, any time during this six-week window, can sometimes be enough to set a kitten along a path that will, if things go right subsequently, allow him to become a pet. However, ideally all kittens should be handled gently throughout this period, maximizing their chances of becoming friendly pet cats.[1]

A kitten that does not encounter people until he is nine or ten weeks old will behave very differently from one that does. At about the end of the eighth week of life, a fundamental change takes place in the kitten's fast-growing brain, one that reverses, almost overnight, its reaction to encounters with totally "alien" animals—which would include, most crucially, humans. Whereas a young kitten is both trusting and curious, willing to accept the attentions not only of its mother but also of people, a slightly older kitten will have become wary of novel experiences of anything that moves and is larger than he is. Instinctive responses toward prey animals are not inhibited, but potentially risky encounters with large animals, such as dogs, are actively avoided—most older kittens will simply run away.

Thus, the distinct social behavior of feral and pet cats becomes more or less fixed as kittens enter their third month of life.

This is not to say that kittens stop learning about how to react to people once they are eight weeks old, rather that they require at least some pleasant experiences of being with humans before that time if they are to continue on their path toward becoming a pet. Apart from pedigree cats, which are usually homed at about twelve weeks of age, domestic kittens generally move to their new home at the beginning of their third month, at a time when their brains are still growing fast. The experiences they have had of people while with their mother should have primed them to interact in a friendly way with their new owners, but far from ceasing to learn at that point, they keep on refining the way they behave over the next several months. They seem to try out different ways of getting their owners' attention and influencing their behavior—for example, working out whether sitting on a person's lap is a pleasurable activity for both parties.

As an illustration of just how flexible their behavior is at this point in their lives, one of the skills that young cats learn during this period is how to get our attention, specifically by meowing. This vocalization is so characteristic of cats that it is often presumed to be instinctive—indeed, in Chinese it forms the word for "cat"—*māo*. However, feral cat colonies are generally rather silent places, the cats communicating mainly by body language and scent, and not vocally: cats generally reserve this sound for communicating with humans. Young kittens instinctively meow to their mother to attract her attention, but she stops responding when she wants to wean them, and when they discover that calling out to her no longer provokes a reaction, they stop doing it. However, the meow, though dormant, evidently remains in their repertoire, and as they adapt to their new home, cats find that meowing has become effective once more, this time in attracting the attention of our own species. In this way they overcome our habit of fixing our attention on a book, a TV or a computer screen and thus failing to look up as cats enter the room, as they expect we should. Some cats simply learn a single all-purpose meow, while others develop a repertoire of different meows, perhaps one that means "I'm hungry" while another that means "please let me out." A recent study by Sarah and her colleagues demonstrated just this. She recorded the meow vocalizations of cats in a variety of different contexts that included during preparation of food, affectionate interaction with the owner and when stuck in a different room from their owner. Owners then listened to these vo-calizations, which comprised both vocalizations recorded from their own cat and some from a cat that was unknown to them, played back in an order unknown to them. We found that owners were bet-ter at identifying the contexts in which the vocalizations were pro-duced by their own cat than they were for vocalizations produced by an unknown cat. Ultimately, every cat works out for himself which meow produces the desired result, with the consequence that each owner and each cat build up a private "language" between them that both understand perfectly but means little to any other owner.[2]

Although the process is poorly understood, young cats make many other changes to the way they behave in their attempts to adapt to life in their new households. Certainly by the time they are two years old, cats have developed their own individual ways of in-teracting with their owners, ways that bear little resemblance to how they behaved at four months or indeed to the way that their two-year-old brothers or sisters in other homes would behave in the same situation.

Each cat's idea of what constitutes a human being will be slightly different, and this will in turn influence what he learns during adoles-cence. Kittens do not seem to bond specifically to individual people: rather, each kitten builds up its own composite picture of the human race—how comprehensive this is will depend on how varied his early experiences were. Thus, cats that have encountered only women during their formative weeks may well have difficulties later in learn-ing how to trust men or children. It is almost as if, and may actually be, that they do not classify small humans, or humans with low-pitched rather than high-pitched voices, as "humans" at all. Rather, they may have to construct a brand-new category for each. Ideally, all kittens should have been gently introduced to women, men and children before they are eight weeks old (rough play—or worse—can have the opposite effect and may even lead to a lifelong aversion). It's often recommended that the kitten be exposed to at least four persons, but variety is probably more important than the precise number. Variety should not be restricted to age and gender but should take account of appearance too—people come in all shapes and sizes, and large amounts of facial hair or religious dress can hide parts of the face and head. Cats will have the best chance of categorizing all as people later in life if they have a wide exposure to different appearances of people as kittens. Ideally, each litter should be handled for an hour a day, broken up into several short sessions, making sure that the most timid individuals get their fair share of handling.[3] Kittens that miss out on some of these important experiences may be able to "catch up" in the first month or two after homing, but others may never learn to fully trust men or children (for example) without additional help. Luckily, this is one problem that training can help with.

THREE FACTORS COMBINE TO INFLUENCE HOW ADULT CATS BEHAVE toward people in general. The first is the personality that they were born with: some kittens are genetically more timid than others, more inclined to recoil from anything novel. Second is the range and nature of experiences encountered between two and eight weeks of life. Left to their own devices, timid kittens will be less likely to choose to interact with people than their bolder brothers and sisters will be. It is therefore best that the kittens' owner or carer adopts a structured approach to socialization rather than let all the kittens interact with people spontaneously, so that timid kittens are not overlooked. Third, the situations that each cat encounters during the next few months of his life will have further changed the way he behaves toward people. If a young cat has unpleasant experiences—for example, being repeatedly teased by children—he may become wary of children. Apart from the initial genetic factor, all of these behaviors are learned and hence can be altered by training. Learning that people can be loved, not loathed, is essential for a pet cat if he is to live contently in a world full of cat lovers. Timid kittens can be taught to be more trusting. Cats that have learned to avoid people of a particular type, or just unfamiliar people in general, can, contrary to much popular opinion, be persuaded to change their mind-set.

For a cat that shows some willingness to interact with people but is a little cautious, finding the least threatening way of greeting that cat will reward any curiosity he shows to people in general, thereby building the first step toward a relationship that is pleasurable for both parties—cat and human. Thus, although a great deal of training involves providing the right consequences when a cat does something, thereby encouraging him to perform that behavior again—for example, stroking the head of a sociable, interactive cat when he jumps into your lap—another important aspect of training is setting up situations that encourage the desired behavior. For example, if you want to encourage your cat to jump onto your lap, make your lap accessible and comfortable and make sure that whatever you do subsequently is perceived as welcoming and inviting to the cat—it's not every cat that likes this type of attention.

When deciding how you should greet a cat, consider how two friendly cats greet one another. Before they approach, they may perform a friendly call, known as a chirrup, before meeting nose to nose and sniffing each other's faces. At this stage, some cats will choose to simply walk away in a relaxed fashion, while others will continue the interaction by rubbing their heads and cheeks against one another. Cats that are particularly friendly with one another will move farther past each other, rubbing the sides of their bodies against one another, often still facing in opposite directions. Such an interaction may finish with their two tails intertwining. During this time, particularly at the initial face-to-face meet, the cats' eyes will appear soft and almond shaped, and they may be seen closing their eyes in a slow, blinking fashion.[4] When you are attempting to greet and stroke a cat, mimicking as many of these actions as possible will increase the chances of a positive outcome, one that the cat learns he would like to repeat, thus building your relationship.

Although it is not possible to reduce our own size to that of a cat, we may appear threatening when we tower above them, so sitting still in a chair or on the floor can encourage interaction. We can mimic the cat's friendly intent, as shown by his eyes, by avoiding staring at him, instead looking past him and closing our eyes slightly or blinking slowly. Gently calling the cat's name can give him some indication that you are friendly before he decides to make contact with you. Cats generally attend well to soft, higher-pitched sounds—recall that their hearing is most sensitive to higher frequencies—so using the voice that we reserve for babies and small children can work well. In addition, we can mimic the chirrup vocalization by forcing air between our top and bottom lip, making a high-pitched "prrrrrp" sound. However unnatural and awkward they may seem to us, such actions are likely to reassure cats that our intentions are amicable, thus encouraging them to choose to approach. While maintaining friendly and inviting body language by mimicking the body language of a friendly cat, it is important that we do not move toward the cat, instead remaining stationary, which allows the cat to voluntarily observe, approach and explore us if he so wishes. Spontaneous movements on the cat's part can be rewarded with softly voiced verbal praise or by gently tossing a treat in the cat's direction (see Key Skill No. 1, reward spontaneous observation and exploration).

If the cat decides to move closer to investigate you, you can make yourself even more approachable by slowly extending a loose fist or finger away from your body, allowing the cat to investigate you safely without having to come too close. Presenting a hand like this gives the cat the opportunity to sniff and rub his face against it if he wishes to, just as he would another cat's head. Utilizing such a hand position also helps us to resist the urge to start stroking the cat straight away—this is important, as the interaction is more likely to be successful if the cat voluntarily chooses for it to continue, rather than having to tolerate a physical intrusion that he may not yet be quite ready for.

If the cat does react positively, for example, by rubbing his face against your hand, you can respond by gently touching his facial

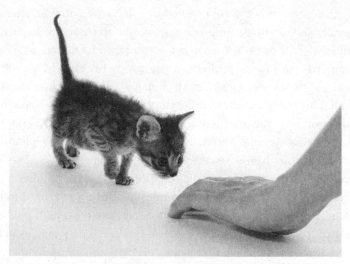

Skippy, an orphaned kitten temporarily fostered by Sarah,
learning that human hands are nothing to fear.

A visitor extending her finger,
giving Cosmos the opportunity to investigate.

area—cats respond best to touch in those parts of the face that contain scent glands: under the chin, on the sides of the lips and behind the whiskers, and the areas in front of the ears where the hair is slightly thinner. Only if the cat responds positively to this by rubbing his head and face against your hand should you then continue the interaction. Cats generally prefer most physical interaction to occur around the back of the head and the sides of the face, so any interaction down the body and tail should be brief and light. It is also best to allow the cat to choose when the interaction will end—this gives the cat a feeling of control over the situation, which is a rewarding experience in its own right and thus will contribute to the probability that the cat will repeat the experience in the future.[5]

Some cats enjoy sitting close to us, even in physical contact. For example, when we are sitting on the sofa they may choose to lie against our legs or on our laps. Only cats that get on extremely well with one another will sleep and rest in physical contact with each other: thus, if your cat chooses to do this with you, it shows he views

Cosmos reacting to a loosely clasped fist by rubbing his face against it.

you as a genuine companion. Just as with the greeting, it is essential to let the cat make his own decision whether to do this, rather than forcing him into whatever position you prefer. Once he has chosen to lie on or next to you, you can reward the behavior with gentle verbal praise, stroking or food, thus encouraging him to do it again.

Some very friendly and sociable cats find being picked up and carried enjoyable. This is not in every cat's nature, because many consider such a situation restricting and difficult to escape from. If you would like a cat that you can pick up and cuddle in this manner, it is important that the cat experiences such interaction from a very young age (ideally starting from that early socialization period of two to eight weeks) and that every instance is accompanied with lots of rewards. Such interactions should be short initially, and cats should always have the opportunity to leave whenever they want.

Skippy has already found that a gentle stroke from a human finger is pleasurable, so this can be used as a reward to teach him that being held off the ground can also be enjoyable.

It is possible to teach cats to ask to be picked up if it is something that they find very rewarding. Usually such cats will spontaneously place their front paws on your legs or body in an attempt to get closer to being picked up. Such behavior can be rewarded with the desired outcome of being picked up. Herbie was particularly sociable and would often climb up onto my chest, inviting me to encircle my arms around him to support him so he could rub his face on mine. Because I responded to his requests (of course I did), he in effect trained *me* to cuddle him in the way he preferred.

At the other end of the social spectrum are cats that are very anxious, timid or fearful around people. This nervousness may relate to the situation—for example, if the cat is new to its surroundings or has had previous negative experiences with people—or it could relate more to his personality. The most important thing we can do in such situations, even though it can feel very unnatural, is to start off by ignoring the cat. This means making no attempt to touch, talk to or even look at him. Although we are often good at not making direct eye contact when we are asked to ignore something, subconsciously we may be tempted to make surreptitious glances in that very direction. However, a timid, fearful or anxious cat can feel threatened when looked at, probably because cats generally stare at one another only during confrontations (they watch each other all the time but usually only out of the corner of their eyes). Thus, by averting our gaze we can help cats to learn that we are not threatening and have no intention of forcing our attentions on them. This way, they should gradually learn that it is safe to remain in our presence, even if this is at a distance. This will begin the process whereby they learn that we are not a threat to them and are not going to demand anything of them.

Although it goes against human nature, it is important that you maintain this stance of pretending the cat is invisible until he learns to feel comfortable in your presence. When he does, he should start to show signs that he is relaxing—for example, pupils that are not dilated, upright ears, and a tail that is hanging loosely and especially not tucked tightly under the body. For some cats, this transition from anxious to feeling comfortable in your presence may take only a few minutes, for others it may take a few days. Once you reach this stage, you can start gently tossing high-value food treats such as small pieces of chicken or ham in his direction. Toss them in a manner that allows them to land just in front or to the side of him and not on him, so as not to startle him. Rewarding his relaxation around you in this manner will help your cat perceive your behavior as harmonious. If he does not eat the treats, this could mean one of two things: either he is just not motivated by food (and may never be), or he is not ready for this level of interaction yet. If the former is the case, try interacting with him using a toy on a very long line, such as a fishing rod toy (if he is particularly timid, you may need to extend the length of the string or the wand that the toy is attached to). If he does not react to this with play or inquisitive investigation, then he is probably not ready, so go back to ignoring him. Cats are inquisitive animals, so his apparent disinterest should not last forever.

However, if he does eat the food treats (or plays with the toy), you should notice that while doing this he is gradually getting closer to you, in anticipation of the food or game—even if initially this may only be a few centimeters nearer. If so, you should reward his decision to come closer with more treats tossed close to him or the toy thrown in his direction for a short game (before putting it out of sight, to keep it interesting). At this stage, it's better not to lay down a trail of treats that lead to you or pull the toy closer to you, because the cat may become distracted and then suddenly realize he is right next to you and panic, as he is not yet comfortable being that close. Rewarding after each little voluntary movement closer to you reinforces the cat's decision to move, which in turn will help him to feel that he is in control at all times: this will help the interaction progress. Try to fit in training sessions of a few minutes' duration a couple of times a day, and at each successive session, withhold the reward until the cat chooses to be a few more centimeters closer to you. This technique "shapes" the cat's behavior, bringing him closer to you, albeit at his pace, thereby keeping him emotionally comfortable with the situation. Once your cat happily approaches you, you can begin to work on greeting behavior (as described above). If the cat is equally nervous with other people, the same training protocol may need to be repeated for other family members, including children, and even for visitors to the home.

THERE IS ONE MEMBER OF THE FAMILY FOR WHOM SUCH A TRAINING protocol will simply not work, mainly because the person is not yet able to fully control his or her own actions. You may have guessed that this person is a newborn baby. Because a baby is very much a permanent resident to the home and not simply a transient visitor, it is vitally important that your cat learns to comfortably live alongside your baby—after all, he or she is going to be there for quite some time! Although we may know that a baby is a little human, cats almost certainly don't: to them, a baby must seem to be a totally different species than its parents. Babies look and—even more important for a cat—smell quite different from adult humans.

Babies behave in very different ways than adults—they are much more spontaneous, inconsistent and unpredictable in their behavior, all traits that many cats find difficult to accept. For example, babies may roll around on the floor and are often very noisy, suddenly gurgling or crying. The curious nature of babies means that when they get a little older, they can show a lot of interest in cats and when at the stage of crawling may be keen to place their faces close to the cat, reach out to grasp or poke him and pursue him if he decides to flee. A crawling baby is approximately the same height as a cat, and thus the cat may suddenly encounter a small person from a much more up-close and personal view than they would an adult. Cats can consider face-to-face encounters threatening, particularly if they are being stared at directly, as babies often do.

To make the situation even more alarming for the cat, along with the introduction of this unusual small person comes a multitude of new and peculiar items—for example, baby seats that vibrate or swing, toys that light up and play tunes, and prams and buggies with large wheels that bring the smells of the outdoors into the home. Changes in smell and the physical setup of the home—not just from these new items but also from all the decorating and furniture moving prior to the baby's arrival—can be daunting for cats. Cats are generally not fond of any form of change, particularly when it occurs rapidly and in an area that the cat considers the core of its territory and thus the place it should feel safest in. Cats that are generally timid may find new items particularly frightening, although a particularly confident cat may view them as his new toys.

The cat will also discover that his routine has been disrupted. The owner's free time will be much more limited once the baby comes home, and daily routines will change, in turn influencing the time that certain events occur in the cat's daily life, such as feeding, play, attention and being allowed outside. Undoubtedly, many of these changes are unavoidable in the preparation for and arrival of a new baby. However, with the right preparation we can go about such changes in a manner that teaches our cats how to cope with them and thus minimizes any anxiety.

Because cats cope best with gradual change, plan any alterations to the home in a staggered fashion, so that at most only a few occur at any one time. For example, if you plan to paint the baby's room, move the furniture out of this room several days before you start any painting, giving your cat time to adjust to each change before introducing another. Reward any spontaneous investigative behavior with tasty treats or the cat's favored reward.

You may decide that when the baby comes you do not want the cat to have access to the nursery. You may also decide that you do not wish your cat to be in contact with certain items, such as the pram and anything else that the baby may sleep in. If this is the case, it is important that you do not allow your cat access to such items prior to the baby's arrival, as it will make it only more difficult to train your cat to ignore them later after he has learned they are warm, relaxing places to be. Thus, although we may wish to reward initial investigation such as sniffing of such things to make sure they are not perceived as frightening, such rewards should be given some distance away—for example, by tossing a treat across the room once

Relaxed investigation of new baby items can be rewarded by tossing a treat away from the object.

the cat has sniffed the pram. This will dissuade the cat from spending time in contact with such items while teaching him to be comfortable in their presence.

Similar considerations apply when the baby comes home. It is important to teach the cat to enjoy the general presence of the baby and his related paraphernalia (because that brings treats) but not to reward him for being in physical proximity or direct contact with the baby and his equipment. This may seem like a subtle distinction, but it is important to avoid having the cat purposely touching the baby or sitting in her chair or cot, thinking that this is going to produce a reward. It is therefore important to provide your cat with additional alternative places to rest and things to play with, such as a bed equally appealing as the baby's pram, one that is warm, enclosed and raised off the ground—an "igloo" or cardboard box with soft blankets work well. Your cat should then be rewarded whenever he chooses to use his own personal spaces and toys, especially in the presence of the baby and her paraphernalia. Although playing with toys should be rewarding in its own right, providing food rewards in the cat's bed area will encourage him to spend much of his time there.

Some of the new items you bring into the home for the baby will create movement or sound, for example mobiles, activity centers and swinging seats: teach your cat to feel comfortable around them before the baby comes home. Do this by introducing them one at a time over a period of several days or weeks using the Key Skill No. 2, systematic desensitization and counterconditioning—this will expose your cat to each item at a pace that will not cause him any fear. Make sure that you reward the cat for relaxed behavior during such exposures. For example, if the item plays sounds or music, switch it on briefly at its lowest setting for just a couple of seconds in the first instance. Monitor your cat's behavior at all times so you can check that he is not finding the experience frightening. If your cat looks relaxed or indifferent, switch off the item and immediately reward him—for example, toss a food treat toward him (but not directly at him) and away from the item, in the same manner that you used for getting the cat accustomed to the baby's sleeping places. If your cat appears uncomfortable in any way with the item, switch it off immediately and enlist the help of someone else, so that the next time it is switched on, you and your cat can be at a greater distance from it, thereby reducing its salience—for example, in another room but with the door open far enough to be able to see and hear it.

The same systematic desensitization and counterconditioning training technique can be used to teach your cat to be comfortable with hearing the sound of a baby's cry before the baby arrives. The Internet is full of freely available audio clips, which should initially be played at low volume and when your cat is relaxed. You can reward continued relaxed behavior through any of your chosen rewards: food, gentle praise, stroking or even play. It is important that the sound is always played at a level where your cat shows no signs of discomfort or distress. Gradually, you can increase the volume over several exposures. If at any stage your cat does appear to be uncomfortable with the sound, stop immediately, wait until he is relaxed once more and play again at a lower volume. Through this process, your cat will learn that the sound of a baby crying is nothing to fear and part of the normal background noise that can simply be ignored.

As well as the introduction of new items, there will likely be changes in daily routines, often starting even before the baby arrives. Cats, however, thrive on routine. During the weeks before the arrival of her baby, it is common for a mother to take some time off from work. Although it is tempting to use some of this time to give your cat more attention than usual, this is inadvisable, because his routine would then be altered to an even greater extent when the baby comes and you are inevitably distracted. Instead, before the baby arrives, try to create set times when you will give your cat the opportunity to interact with you: these may be times when your partner or another family member is also present so that when the baby does come, you can still devote these same times to the cat. If the mother is the main carer for the cat, for example, carrying out feeding, grooming and other duties, it will be useful if these tasks are gradually switched over to another family member before the baby arrives. This way, the predictability and consistency that the cat thrives on will be maintained. Also, encouraging independence through the use of puzzle feeders and self-play toys will help keep your cat physically active and in a positive mood while you have less time to play with him in the early stages of caring for a baby.

The alterations to the household associated with a new baby— the new items, the decorating of the house and the baby—all cause changes to the smell of your home, potentially masking the cat's own contributions that he has deposited by facial rubbing on furniture, door frames and wall corners. It is important to re-establish this familiar scent: doing so will help your cat feel safer and more secure during such changes. This can be achieved through collecting your cat's scent (Key Skill No. 7) and rubbing it over new items and newly decorated areas. Recall, scent can be collected by wearing a cotton glove and stroking the cat's facial gland areas with this gloved hand. If you do not have a glove, you could hold any piece of light cotton cloth—such as a clean handkerchief—in the same hand, ensuring the cloth repeatedly touches the facial gland areas. As your cat detects his scent on the new baby items, he should not feel perturbed by their presence and may even further contribute to the scent by facial rubbing on them himself. We cannot smell such odors, as our sense of smell is not as good as the cat's, and there is therefore no need to worry that you will "dirty" the new items. If you have more than one cat, it is important that you do this for every cat, as the scent in the home will be shared, made up of contributions from each cat.

Herbie scent-marking the corner of the
new cot after I have wiped his facial scent on it.

Because cats are such odor-oriented animals, it is also important that we habituate them to some of the new smells that a new baby will bring. Before the birth, start to try out some of the nappy creams and other baby products on your own skin, to allow your cat to become habituated to such smells. Your cat will thereby learn that such smells have no consequences for him and will accordingly find them less disruptive when the baby comes. One smell we cannot recreate before the baby arrives home is the smell of the baby. If you have friends with babies, it is a good idea to ask to borrow a blanket or clothing their baby has had contact with that you can allow your cat to sniff, remembering to reward positive or indifferent behavior toward it. Indifferent behavior is just as important, as we do not want to create a situation where the cat learns that all rewards come from being in close contact with the baby. You may also be in a situation where your partner is able to bring home a blanket or clothing from the hospital with your own baby's odor and allow your cat to investigate it, just before you come home with the baby.

With lots of change going on, it is important that your cat still perceives all his resources to be safe and accessible. If your cat lives as part of a multicat household, the disruption your cat feels from the arrival of a new baby can bring out tension between cats, even those who previously appeared to tolerate one another. Thus, it is a good idea to provide additional resources for all your cats, such as more beds, feeding places (but not more food!) and litter boxes— using covered boxes may also help enhance your cats' privacy and hence their perception of security. If some of these resources are currently in places that a crawling baby will be able to access, consider adding more in places that will be out of reach when that time comes.

Likewise, it can be helpful to provide additional elevated vantage points from which your cat can survey what is going on, such as shelves, and allow him access to windowsills. Later on, a baby gate he can climb over, or one with an integral cat flap, will allow your cat to choose when to be around the baby: again, this should help him learn to feel comfortable with the new situation. This is particularly important for cats that are generally nervous of people, as a new baby tends to bring a lot of visitors to the home. By letting your cat choose whether to interact with them, he will feel more comfortable and, as a result, actually be more likely to interact. Although such additions may not feel like formal training, they will help your cat learn that his things and his space are not under threat of invasion.

AS BABIES GET OLDER, THEY BECOME MUCH MORE AGILE AND more mobile. It is important to prepare your cat for this. Babies and small children should never be left unsupervised with cats, and as soon as they reach an age when they can understand, children should be taught how to interact appropriately with cats. Nevertheless, there undoubtedly will be times when your child reaches out to touch your cat in an inquisitive and somewhat clumsy manner. To prepare for this, you can teach your cat that gentle poking and grasping of his fur and body in the manner that youngsters may do is not something to be feared: for example, you could try wrapping your fingers around his tail or touching with one outstretched finger. Start with very gentle touch in this manner, and immediately after the touch, reward. If your cat is worried by this, you can even begin giving the reward at the same time as the touch, thus using the reward as a distraction from the touch. Once your cat is comfortable with this, move the timing of the delivery of the reward to just after the touch—the touch therefore becomes predictive of the reward. Over time, you should increase the pressure of your touch.

You will also need to generalize your cat's acceptance of such types of touch by encouraging others in the household to do the same. Such touch should involve only a small proportion of the cat's daily total physical interaction with people. By doing so, your cat is less likely to feel overly stressed with slightly more unpredictable or strange handling by a child, such as the occasional small hand grasping his tail or fur—nor become fearful of future interactions. Of course, older children should also be taught the correct way to handle cats gently.

CATS ARE SUCH POPULAR PETS NOWADAYS THAT IT'S EASY TO overlook the difficulties that many of them have in interacting with people. It's a skill that each cat has to learn for himself, and some find this harder than do others, perhaps because they are naturally

Reuben and Cosmos learning about
each other in a safe, calm and controlled manner.

timid or because they were not treated as kindly as they should have been when they were kittens. Because cats have a reputation for being somewhat self-absorbed and willful, it's easy to presume that nothing can be done for the cat that disappears at the first sight of someone he doesn't know. Moreover, such behavior is self-perpetuating, because the cat will probably never get to know people that he has chosen to avoid—hence the pleas that we hear so often, along the lines of "I want my boyfriend to move in with me but my cat hates him—what should I do?" As we've seen in this chapter, this is just the kind of problem that simple training can solve (provided the boyfriend is willing to cooperate, of course).

CHAPTER 5

Cats and Other Cats

THERE IS NO DOUBTING THAT CATS HAVE A HUGE APPEAL nowadays, so much so that many owners want—and get—more than one. It is not uncommon for cats to live in households with one, two or even more cats. Many people presume that all cats will be perfectly happy in one another's company. Sadly, however, it is common for cats to experience stress with this living arrangement—some demonstrate their dislike at having to share their territory with another feline by spending large parts of the day out of the home; some even move out completely. Others attempt to defend what they perceive as "their" territory, categorizing their feline "companion" as an intruder. Such defense ranges from subtle behavior like blocking their feline housemate's access to key resources, such as litter boxes and food bowls, to the more obvious and often harder to stop behaviors—catfights and urine-spraying, both of which can cause considerable nuisance and distress to the owner as well as to the cats. Fortunately, training cats to get along with each other from day one is much easier than trying to turn an established loathing into lasting love.

Cats owe their somewhat primitive social skills to their ancestor, the wildcat. Unlike the highly sociable wolf from which the domestic dog sprang, cats are descended from a territorial, solitary animal that had little reason to evolve the behavioral tools needed to achieve year-round coexistence with other cats. As wildcats changed their habits so they could exploit the opportunities that humankind

115

provided—concentrations of prey greater than anything they would have encountered in the wild, alongside the occasional handout—their solitary natures would have become a hindrance. In the wild, predators that live too close to one another run the risk of overexploiting their prey and ultimately running out of food, which is why a secure territory is so important to them. Yet when we first started to live in towns, prey suddenly became so abundant that there was enough for several cats all in one confined area. Their territorial behavior became a handicap: the advantage would have been with those cats that could concentrate on hunting instead of constantly watching out for rivals. Nevertheless, to this day cats prefer to be solitary when they are actually out hunting—a mouse makes a meal for one cat but is not worth sharing. Although often surprising news to owners, their cat's preference to eat without other feline company, even when faced with enough food to feed several cats, stems from their solitary hunting preferences.

As domestication progressed, cats had to become more tolerant of other cats, but their adaptations to their new environment did not stop there. The males remained solitary, while the females evolved a simple type of cooperation for raising and protecting their kittens, one that is still practiced by farm cats today (and some breeders of pedigree cats also witness it). When food is plentiful, mother cats allow their female offspring from one year to continue to share the core of their territory and to breed there the next: nursing and provisioning of the next generation of kittens is often shared between mother and daughter(s), irrespective of which one was the biological mother. Although few people today have been lucky enough to witness such enchanting behavior, its legacy is evident, even in pet cats, in the generally stronger bond that exists between cats that have grown up together, as compared to cats that first met as adults.[1]

Today's pet cats are also more tolerant of one another because most are neutered. This is particularly evident with males, who are generally intolerant of one another if left sexually intact. In cat colonies, male kittens are driven away by their mothers—and any adult males nearby—as soon as they show signs of the hormone-driven changes that turn them from dependent juveniles to competitive tomcats. Castration stops these changes from happening, such that males neutered before they are six months old behave much like neutered females. Thus, while in a breeding colony the only long-lasting bonds are those between related females, in a household two (neutered) male kittens from the same litter are just as likely to remain friends as two females—or indeed one of each sex.

Cats that get along well will demonstrate and bolster their affection in a couple of ways. They will usually raise their tails upright when they catch sight of each other, although cats that have lived together for a long time may dispense with this nicety. Then, they may either rub against one another—some restrict this just to heads, others venture to include flanks or tails—or settle down to rest in physical contact. While resting, they may groom one another, especially just behind the ears: although this undoubtedly helps to keep both cats' fur clean in a hard-to-reach place, it also has a social significance, reinforcing the bond between the two cats.

Although some multicat households happen to mimic the natural situation by being composed of related cats that have grown up together, the majority do not. Tensions are almost inevitable when several unrelated cats live under the same roof and have to compete for the same set of resources, particularly so when the cats have been introduced to one another as adults. It simply goes against their nature to share sleeping places, food bowls and litter boxes, and although a few succeed in forging a friendship with one of the other cats in the house, many do not.

The cat's origins as a solitary and territorial animal provide another unfortunate legacy: not only are they not inclined to "think social," as a dog might, but their ability to communicate any friendly intentions they might have toward another unfamiliar cat is somewhat limited. Suiting a highly competitive animal, cats have a varied repertoire of hisses, growls, yowls and aggressive postures that signify intent to attack when challenged. They can also adopt a defensive posture, turning sideways and making their fur stand on end in an attempt to make themselves look bigger than they really are—a posture often used in illustrations of cats that depict Halloween. But cats lack the expressive faces that dogs use to communicate with one another in a remarkably nuanced way. Most significantly, cats lack a clear signal that means "I'm not threatening you," or at least, one that can be performed toward a stranger: the tail-up signal conveys this message but is only used by cats that are already on good terms. As a consequence, encounters between cats who don't know one another tend to develop rapidly into a standoff in which neither dares to back down, simply because to do so would invite a chase and the risk of getting attacked from behind. Eventually, one or the other will slink away, constantly looking over his shoulder until he feels he is out of danger, or will resort to suddenly spinning round and then running as if his life depended on it.

Thus a cat's natural response to encountering another cat is to perceive the other as a rival and either flee or attack. And unlike dogs, cats do not usually turn to their owners for guidance as to how to react. Thus they are probably completely unaware of the distinction between a cat that has elected to invade the house from next door and a new cat that has just been brought home by a well-intentioned owner providing a "playmate." They both are, in the resident cat's eyes, simply intruders.

If the other cat is indeed an invader from elsewhere, the owner will probably be almost as keen to discourage it as the resident cat is—especially if the intrusions lead to one cat or the other (or both) spraying urine around the disputed area, for example around the cat flap. Depending upon just how determined the stranger is, the situation may therefore be resolved in the resident cat's favor—although sometimes clashes of this kind can rumble on for months, perhaps confined to the small hours when the owner is asleep.

If the "intruder" is actually a new cohabitee chosen by the owner, both cats will have to adjust to the new regime, and it is here that training can make all the difference between harmony and discord (training is less easy to apply when the other cat belongs to a neighbor, unless the neighbor is prepared to cooperate). Left to their own devices, the two cats will probably skirmish for a while and then divide up the house between them in such a way that each has at least one or two areas where he can rest without fear of being ambushed. An observant and sympathetic owner may accept this as a reasonable modus vivendi and provide each cat individually with what it needs: food bowls and litter boxes in different parts of the house and several well-enclosed and well-separated sleeping boxes where each cat can relax well away from the other. Simply doubling the amount of food provided and cleaning the single litter box twice as often is a recipe for dissension: one cat (often but not necessarily the original resident) will instinctively attempt to monopolize both.

Add one or two more cats to the mix, and the likelihood is that at least one of the relationships between them will disintegrate into open warfare. Rare is the indoor/outdoor multicat household where at least one of the cats is not living part-time elsewhere. However, it is by no means inevitable that two or three cats cannot live in harmony, provided they were introduced to each other in the right way and had suitable personalities to begin with. As with so many cat-related issues, planning and patience are essential, but owners who apply them properly can reap the rewards of a relaxed and sociable household for felines and humans alike.

THE INITIAL STAGES OF PLANNING FOR A FELINE ADDITION TO the household involve giving careful consideration as to whether the current resident cat or cats will be likely to accept another into the home, and whether the house itself is suitable to support a new addition comfortably. There are several pros and cons to consider regarding the introduction of a new cat into the home—and for each case, those pros and cons will be slightly different. Our hearts often overrule our heads when it comes to animals, but in situations such as these, objectivity is the best starting point for creating feline harmony. Only if the pros outweigh the cons, not necessarily in number but in weight of impact they might have, should the thought of another cat actually be put into practice.

Careful deliberation should first be given to whether your current cat (or cats) like other cats. You may feel you have a good idea of this if your cat has previously (and happily) lived alongside another. However, just like people, simply because your cat used to get along with one particular cat does not mean he will like any cat. We know from studies of people interacting with kittens that it takes handling from at least four or five different people during the early period of life (two to eight weeks), when a cat is most receptive to learning about social situations, in order to generalize his knowledge that all people are friendly, not just the specific people who handled him during this time. Unfortunately, there are no corresponding studies with other cats rather than people, but it is reasonable to assume that a kitten that has had amicable experiences with several cats in its early life will be more disposed to view other cats positively later in life. And beyond your cat's early history with other cats, there are several other factors to consider, all of which may have some impact on a cat's ability to view other cats positively.[2]

Taking into consideration the social behavior of free-ranging cats, we now know that pet cats are most likely to get along with each other if they are neutered, young (less than one year), related, and come from parents who show friendly behavior toward other cats. The fewer of these that apply to your current cat, the less likely he will take positively to a new cat. Taking all these factors into account, the situation that is most likely to lead to a happy multicat household is one where two kittens are introduced to the home at the same time, ideally but not necessarily kittens from the same litter. However, if you already own one cat, introducing a kitten to an adult cat is likely to be easier than introducing two adults to each other, although the latter may be possible if the two adult cats are related. In some cases, if the home is big enough, introducing two new kittens, as opposed to one, to a resident cat may be more appropriate, as the kittens will often play together and therefore pester the resident cat less than a single kitten would.

If you are considering obtaining a kitten, be proactive in finding out about the kitten's parents. For many kittens in shelters, such information is not known, particularly with regards to the father, but for kittens purposely bred in the home, such information should be easily obtained. When this information is available, prioritize a kitten from parents that are confident and friendly around other cats: such tendencies are at least partly inherited. When selecting puppies, owners are encouraged to view the mother and (where possible) even meet the dad, but this is something rarely considered when selecting a kitten. If you are obtaining a kitten from a breeder who does not own the father, ask the breeder to put you in touch with his owner so you can view him or at least ask some questions about his temperament and attitude toward other cats.[3]

If you are obtaining a kitten from a shelter, the kitten may already be separated from his mother by the time you are able to view him. Ask about the mother, whether you can meet her and what her response has been to being in the vicinity of other cats while at the shelter. Ask about any previous history they may have on her, such as whether she lived happily with other cats before arriving at the shelter. Recall, cats that have positive relationships with one another will be seen to rest and sleep together, happily share their litter boxes and food bowls, groom one another, play with one another, and rub their faces, bodies and even tails against each other. When considering the gender of cat, an existing resident male cat may find a female cat a less threatening addition than another male, particularly if the age at which the resident cat was neutered was after six months or is unknown. Males who have been neutered as an adult, however placid, may find it harder to live with other cats, particularly other males. Those who have been neutered at six months or earlier (before testosterone had time to influence the growth of their brain) are generally friendlier to one another.

Once you've thought about how your resident cat's age, gender, neutering status and parentage might equip him for dealing with another cat, you should give equal thought to your current cat's early social experiences as a kitten, as these will very much shape how he views other cats. Knowing how much positive social experience your cat has had with both related and unrelated cats in the early formative period of his life, particularly during the sensitive period of two to eight weeks, will give some indication of the likelihood of positive relationships being formed with new cats.

It is important to consider not only whom your cat met during this time but what form the interactions took, how frequent they were and what environment they took place in. For example, consider the scenario of a shy kitten who was the smaller of a litter of two and whose stronger brother displaced him from food and played slightly more roughly than could the little kitten. Then imagine that an intruder cat lives nearby, and even though the owner of the kittens has locked the cat flap to prevent the previously intruding cat from coming indoors again, the mother of the kitten catches sight of this cat through the window and reacts defensively, hissing and spitting—feeling more protective than normal due to the recent addition to her family. The intruder cat will have been the kitten's only experience of an unrelated cat (albeit through a window), and this experience involved a hissing, scary-looking fluffed up creature. He will now be less likely to view cats positively when meeting them later in life.

Compare this scenario to the kitten who was reared in a home with a couple of breeding females who both had litters at the same time. The kitten has not only his own siblings to interact with but also a whole other litter of kittens and their mother, to which he is unrelated. If all get along well, there is much opportunity for play, being groomed by other cats and falling asleep together. Based on this information alone, the kitten from the latter scenario would have the greater probability of getting along with another cat in his adult life. The greater the number of positive experiences with a range of other cats (related/unrelated, adult/kitten) during these early weeks of kittenhood, the greater the chances of the cat accepting other cats as an adult. Remember, however, that it takes two to tango: the same must be true for the new cat to ensure that he has a positive attitude toward your existing cat.

Unfortunately, such a good early foundation to social relationships can be knocked off kilter by later negative experiences with other cats. Cats, like other animals and indeed ourselves, tend to remember negative experiences better than positive ones: the mammalian brain is geared to be more sensitive to nasty events so as to be able to recognize any potential danger the instant it occurs. Thus, any negative experiences a cat has with another cat are unlikely to be forgotten and will shape future interactions, not just with that particular cat but also with other cats it meets.

If your cat has a history of being involved in fights with other cats, it is important to work out whether your cat was usually the victim or the perpetrator. A perpetrator learns that aggression works to keep other cats at bay and thus will probably repeat this tactic if forced to share his home with another cat. A victim will likely feel nervous and on edge around other cats and need a lot of confidence-boosting support to be able to accept another cat, if he can at all. For such a cat, the introduction process should be carried out as slowly as possible. Thus, cats with no negative experiences with other cats (current or previous, belonging to the same household or those in the neighborhood) stand a better chance of happily living alongside a new cat than those who have a history of negative experiences.

Finally, it is worth giving consideration to how your cat currently interacts with any cats he encounters—these could be other cats he currently lives with or cats he may meet by chance (for example, if he has outdoor access he may encounter neighboring cats). If he has ever been in a boarding cattery, visited the vet or been entered for a cat show, consider how he reacted to the sight and smell of other cats around him. Although the latter scenarios may involve some elements of distress unconnected to feline rivalry—for example, being taken away from his familiar territory—a highly negative response directed toward other cats, whether hissing, growling or spitting at them, is a clear indication that bringing a new cat home is unlikely to be viewed favorably by your cat. If your cat simply ignores other cats, however, this does not need to be considered a negative outcome—in the face of an unknown cat, ignoring it suggests your cat does not feel unduly threatened, and this provides a good starting point for the introduction of a new cat.

After thorough contemplation of the characteristics of your current cat(s)—in summary, his previous encounters with other cats and his current behavior toward other cats—you should have a good idea whether he is likely to accept a new cat. Unfortunately, owners whose existing cats were obtained as strays or as adults from shelters are highly unlikely to have this information, nor will they have any way of finding out. In such cases, the decision is often much more difficult to make as there is less information to base the decision on. The less information you have, the higher the risk of an unsuccessful introduction to a new cat.

Once the suitability of your cat to live alongside another has been deliberated, you are halfway toward making the decision whether to obtain another cat. The next step of the decision-making process is considering whether your home can support another cat in terms of the resources you can provide, their distribution, and how much time you can devote to each cat.

Cats can be regarded as the ultimate control freaks. They like to have 24/7 access to all the resources they need in life—feeding bowls, water bowls, puzzle feeders, toys, beds, hiding places such as tunnels and boxes, vantage points such as perches and shelves, litter boxes and scratching posts or pads (some cats prefer to scratch in a horizontal position rather than the traditional vertical position—a scratching pad placed on the floor allows this). Forming an orderly queue does not come easily to cats. Furthermore, sharing the same litter box or sleeping in the same bed (even if at different times) applies only to individuals who consider themselves part of the same social group. The current cat(s) are unlikely to consider any new cat as part of their social group, at least initially, and thus enough sets of resources must be provided so that no cat is forced to share.

If your answer to any of the following questions is no, then the likelihood of your current cat(s) accepting a new cat is diminished. The more you answer no, the less likely your resident cat will have the opportunity to learn that the new cat is not a threat.

- Can I provide as many of each resource as there will be cats in the household?
- Can I distribute each resource throughout the house, so that each one is well separated from all the others?
- Can I place resources in a manner that one cat cannot block another's entry or exit to that resource? For example, can I avoid placing resources in corners of rooms and provide hiding places, such as cardboard boxes, with both entry and exit holes?
- Can I give each cat individual attention in a form he enjoys (e.g., play, stroking, grooming) away from the other cats?
- Can I provide a room dedicated entirely to the new cat for the initial introduction period?
- Can I construct a barrier between the room that will be dedicated to the new cat and the rest of the home, thus allowing the cats to be physically separated but still able to see and smell one another? (If a barrier is not possible, you might use a large crate instead.)

If you come to the conclusion that there is a good probability that your cat could cope with the addition of another, and it is possible to have adequate resources in place within the home, then the next step is the introduction process. The initial introduction of cats is often crucial in determining their future relationship. Therefore, it should never be rushed.

The first step is to create a safe, secure area in your home where the new cat will initially be based exclusively. This should be a room that your current cat does not often use, for example, a spare bedroom or study. You should place all the items your new cat will need (litter box, food bowl, water bowl, scratching post, bed and toys) in this room, not too close together, and shut the door. If your new cat does not come with his own belongings, purchase new items: do not "borrow" some of your current cat's belongings as they will smell of your cat, which may scare the new cat. Also, you risk upsetting your existing cat by depriving him of some of his cherished "possessions."

Introducing your cat to a new home

Cats are usually at least as strongly attached to the place they live as they are to the people who live there, and therefore they hate it when their owners move house. Transported out of their familiar surroundings, they do their utmost to get back to the place they regard as "home." However strongly attached they may be to their owners, they miss their old haunts far more, at least for the first fortnight or so—hence the advice not to let a cat go outdoors for two or three weeks after a house move. As the cat comes to feel more secure in his new surroundings, so the lure of his old neighborhood gradually diminishes.

The creation of a single safe room suggested here when introducing a new cat to a household can also be used when moving house with your cat—by initially confining your cat to one room in the new house containing all his familiar belongings, he will be surrounded by his own scent and thus feel safer and more secure in this initially alien environment than he would if he was faced with an entirely new house. Using the technique you would use to get two cats accustomed to the scent of each other, you can collect your own cat's scent and impregnate his new home with it, rubbing it onto furniture and edges of walls at cat-head height prior to your cat being allowed into each room. Thus, as you gradually introduce your cat into different rooms within his new house, your cat should detect his own scent and feel less threatened by all the change.

If possible, set up the room at least several days before the new cat is due to come home, so that your existing cat can learn that this is a room he can no longer enter. If he goes to sniff, paw or scratch at the closed door, simply ignore this behavior, and if it does not stop, entice him away with a game with a wand toy. The resident cat must not associate that part of the house with negative feelings before the new cat has even arrived![4]

Care is needed when bringing the new cat to your home for the first time. When the new cat arrives, take him to his new room, taking care that your current cat does not see him. Place the new cat's carrier in the corner of the room or on a raised surface such as a bed, and open its door. Do not try to take your new cat out of his carrier as it is better if he learns that coming out on his own is a safe thing to do: that way he won't feel at all forced but instead he'll feel in control of the situation. This will enhance his security and ultimately help him to learn. You can, however, talk to your new cat in a calm, soft voice, from a short distance away. If he pokes his head out of the carrier, reward this with gentle praise and a food treat (although he may not want to eat until he feels more settled).

If he is showing no signs of coming out of his carrier, do not worry. He may need some quiet time to settle down and learn that there is nothing to fear in this new room. Although dogs can benefit from human reassurance in times of uncertainty, cats can find it disruptive or even distressing. Thus, if your cat shows no signs of coming out of his cat carrier, leave the room quietly to allow him to come out in his own time and to explore his new surroundings. He may not do this until everyone in the household has gone to bed, so make sure he has plenty of food and water. Carry on with your normal routine with your existing cat. Any change in your behavior will only alert him that something is amiss and put him on edge.

The introduction process should involve familiarizing the two cats with one another gradually, one sense at a time. Thus, we need to consider hearing, smell, vision and touch—in that order. This staged process, known as systematic desensitization (part of Key Skill No. 2), prevents the experience from being overwhelming and allows both cats to learn slowly, processing each piece of information individually and feeling comfortable with it before building up to the next stage. As humans, when we meet people for the first time, we tend to see them approaching first, we then introduce ourselves verbally and only then do we shake hands. It may take several meetings before we are comfortable enough to touch one another more than this—for example, hugging, patting on the back or kissing on the cheek. Imagine a situation where, before even saying anything to you, a stranger walked up to you and hugged you. Rushing the introduction process for cats can create similar misapprehension and tension. Although vision is one of our most utilized senses and talking our greatest means of communication, the cat's primary method of navigating its environment and communicating with others is through its sense of smell and its ability to deposit scent from glands in the body, such as those in the cheeks used when rubbing.[5]

The first place to start when introducing cats is to allow them to smell one another before letting them make visual contact. We may not have been able to completely control what they may have already heard, such as meows for food or the other cat walking past on the opposite side of the door. However, we can create relatively controlled introductions to each other's scent. Only begin this stage once both cats appear in their normal routine—settled and relaxed in their individual parts of your home.

To introduce each cat to the other's scent, we first need to collect the scent on an item we can give to the other cat. This could be a light cotton glove put on specifically while stroking the cat or some hair from a grooming brush or just a small piece of cloth that the cat has laid on: whichever, the method used to collect scent from the cat must be pleasurable to the cat (and therefore may be different for different cats). Ideally, the scent should be collected from the area around the scent glands on the face, which are under the chin, in front of the ears and behind the whiskers: the more times these areas touch your method of collection (glove, brush, cloth), the greater the concentration of scent captured (see Key Skill No. 7 for full details). However, to begin with, create a weak concentration of scent by stroking or brushing only a few times or leaving the cloth in the cat's bed overnight. More concentrated samples can be used after the cats have adjusted to the weaker scent. Once you have collected a scented item for each cat, place each in the other cat's environment—anywhere on the floor where he is likely to encounter it.

Watch each cat's reaction to the other's scent. This will give you an indication of how likely the cats are to accept one another. If either cat hisses at the scented item, this suggests he finds the mere scent of another cat threatening, and the introduction process should be carried out even more slowly and carefully than usual. If either cat refuses to go near the scented item, this is also an indication that the cat is struggling to accept the scent of the other, presumably interpreting the odor as an unacceptable invasion of his territory. In either case, remove the scented item and try again in a few days. The cats should come to accept the cloth, glove or brush over several repetitions, as they gradually learn that the smell has no further consequences to them. If the cat sniffs the scented item and appears indifferent, we have success—the cat has learned that the smell of another cat in his environment brings no great consequences. If the cats play with the cloths, brushes or gloves, or rest or sleep on them, even better.

I have used this process many times over to integrate cats into groups of up to four in a cattery setting. These were cats that had been relinquished by their owners or by animal welfare charities that had no available space within their own catteries. These cats stayed with us at the university until we could find them homes, giving us the opportunity to study their behavior and correct any problematic behavior. I had many more successful introductions than not (in fact, there was only one cat I could not successfully integrate, and consequently he was rehomed as a single cat). However, there were a few cats who showed avoidance behavior toward the object containing a new cat's scent. For these few cats, the subsequent introduction process always took longer than it did for the majority of cats. I recall one particular cat, Caragh: on the first introduction to her pen of another cat's scent on a small cloth, I had switched on the pen video cameras to watch her behavior remotely. After initially sniffing the cloth, she quite spectacularly hissed at it, despite there being no cat to be seen. However, over repeated exposures to the scent, the hissing subsided, and I progressed to gently introducing the cat from which I'd taken the scent, first in an adjacent pen and then directly into Caragh's pen. Although there was never any overt aggression, it was clear Caragh was not happy with this arrangement—she avoided the new resident at all costs and spent more time in her bed. The new resident was therefore removed back to his previous pen. My initial thoughts were that Caragh would only ever be happy on her own and that she did not enjoy the company of other cats in general. However, circumstances led me to try her again a few weeks later, with bedding scented by another cat. This time there was no hissing; instead, within hours I found her curled up asleep on this bedding. I therefore felt it was safe to try an introduction with the owner of this bedding: the sub-sequent introduction went smoothly, and the cats happily shared a pen throughout their stay. Without a previous history for Caragh, who knows what was influencing her to object to the company of one cat but quite clearly enjoy the company of another? The point of the story is that the initial response to a scented item may give clues as to how the next stages of the introduction will go, so observe carefully.

Cosmos remains relaxed while smelling Herbie's scent on a glove.

THE NEXT STEP IS TO PLACE A SCENTED ITEM WITH A GREATER concentration of the other cat's scent in each cat's environment. If the cats do not respond positively, as before remove the items and try again in a few days. If the cats do respond positively, move the items to the cats' beds to allow them to sleep on the items, thus creating items that mix the scents of both cats—creating a communal scent, rather like the scent shared by members of a cat colony. You may find that the two cats do not accept each other's scent at precisely the same moment; this is absolutely fine—just work at the more reluctant cat's pace, which is possible at this stage, as they are not meeting face to face.

Once both cats are comfortable with having the other's scent in their part of the home, it is time to introduce their mingled scent to physical objects within their separate environments. By doing this, we create the illusion that the two cats already seem familiar to one another, have mixed their scents and thus are part of the same social group (even though at this point they have never actually met). Spread this communal scent around their separate environments by rubbing the scented items on furniture and corners of walls at cat-head height. For some cats, this process can be completed in a few days, but for others it may take weeks.

Later, when both cats appear relaxed in the presence of the mixed scent, it is finally time to allow them to see one another. To prevent the cats from interacting physically at this stage, create a barrier either in the doorway to the new cat's room or between doorways separating two distinct areas of your house. If you decide on the latter, give your new cat time to explore the newly exposed areas of the home before coming eye to eye with your resident cat. A barrier can be made by installing a mesh door or screen or by fitting two baby gates (initially covered with mesh and secured to each other to block wandering paws) one on top of another. It is also a good idea to make part of the screen completely solid (for example, by tying a piece of cardboard to it) to allow either cat to get out of the other's view quickly if he is feeling in any way threatened. While the screen is being fitted, it is a good idea to confine the new cat to his cat carrier and your resident cat to another room, so neither can escape or have an impromptu meeting. Involving another person in the process, so that one person is with each cat, can help greatly.

If a barrier across a doorway is entirely impossible in your home, the new cat can be placed in a large dog crate (not one that has previously been used by dogs!). However, the cat must have had previous positive experience of the crate, through having had free access to it within his room, placing comfortable bedding and treats within it to help create positive emotions. The cat should enter the crate voluntarily and should never be pushed nor should the door be shut immediately behind him. Part of the crate should always be covered with a blanket to enhance the cat's perceived security, and there should, ideally, be additional hiding places within, such as a cardboard box or igloo-style bed. If the cat finds being in the crate aversive, the introduction will not work: the new cat will already be in a negative mood, which will bias his perception of other things he encounters at this time. Even worse, he may come to associate the presence of the resident cat with his negative perception of being in the crate. Therefore, before any introductions, care should be taken to make sure that the new cat is fully comfortable with being in the crate. Because of the necessity for this additional training, barriers are often simpler to use than crates.

The moment when you allow the cats to meet, separated only by the barrier (or the crate), needs careful preparation. Ideally, choose an occasion when something really positive is occupying their attention, for example, during their meals or when you are playing with them. We want to reduce the likelihood that the two cats will end up staring at one another, which each can perceive as aggressive: thus, a situation that encourages fleeting glances is best. Feed or play with each cat as far away from the barrier as possible. Scattering dried food or feeding with puzzle feeders can increase the time involved in eating, thus prolonging the time the cats are within each other's view but their attention is elsewhere. Recall that feeding is a solitary activity for most cats, so the distance between the two should be as great as possible while still allowing some visual contact with one another. During the course of a meal, cats naturally have short bouts of time where they stop eating. During these breaks, they are likely to glance up at the new cat. The perfect situation is where each cat feels comfortable enough to notice the other but then go back to feeding. If this does not occur, close the door between the cats to obstruct visual view of one another and try again on another day, with the door open and screen attached, but offering a higher-value food to really hold each cat's attention; or try a different distraction, such as playing with toys that stimulate hunting behavior. The presence of food or toys helps to change a potentially wary perception of the other cat into a positive one (the counterconditioning part of Key Skill No. 2).

If at any stage either cat moves toward the barrier (or crate) calmly and in a friendly manner, let the cat continue, and afterward reward such behavior with food treats or gentle praise. At no stage should you force the interaction. If one of the cats evidently feels uncomfortable with the other's presence, gently encourage the more confident cat away using a wand toy or food treat, to create greater distance between the cats. If both cats appear comfortable, just sit quietly and allow the cats to sniff one another through the barrier. You can also rig up two toys on a string that fits through or under the barrier, so that each can play with the toy at his end of the string, thus further teaching your resident cat that the presence of the other cat in his home, far from being a threat, can bring reward. It is important that you and your helper stay calm throughout and that your movements are subtle: try sitting on the floor rather than standing, so that you appear relaxed and are not looming over either cat.

There is another way you can distract your resident cat from looking too intently at the new cat. If your existing cat is particularly human focused, one useful task to teach him is to look at you when asked. Making an interesting sound such as sucking air through your pursed lips is often enough for a cat to stop what it is doing and look at you for a second. If you reward this look—the most effective way is to first mark the behavior (see Key Skill No. 4) and then provide your chosen reward—and repeat the process of making the sound, marking whenever he gazes toward you and subsequently offering a reward several times, you can teach him to look at you every time you make this noise. If your cat does not look at you when you make

Cosmos and Herbie are playing with toys
attached to either end of a string,
each on his own side of the barrier.

this noise, follow the noise with the visual presentation of a food reward, bringing it close to your face (but out of the cat's reach to prevent any swiping) to lure (Key Skill No. 3) the cat's gaze to your face. Once the cat is looking at you, give the cat a reward. After several repetitions you can fade out the use of the lure and your cat should look at you when he hears the noise (without the sight of food), which you should reward with the food. Now that the behavior is being offered reliably, it is important that you do not reward every incidence of eye gaze toward you, to make sure the offering of the behavior is maintained in the long term (see Key Skill No. 8).

Training this "look at me" cue provides you with a way to break your resident cat's gaze toward the new cat. Thus, if he starts staring at the other cat in a way that could be considered threatening, use your "look at me" cue to get him to direct his gaze toward you, and immediately reward him. Not only will the cat you are training learn that looking at the other cat and then to you brings a reward; the other cat will learn that another cat looking at it is not the precursor of a fight, and therefore there is no need to flee. Equally, you might be able teach your new cat the same trick—which will also help to strengthen the newly emerging bond between you and your new cat.

Herbie is demonstrating "look at me"
in the presence of Cosmos.

Repeat the process of allowing the cats to see each other through the screen several times a day over several days. Initially, small and frequent exposures are much better than one long exposure to each other. If all is going well, you can begin to leave the door open (but with the screen still attached) for longer and longer periods, but do increase the duration of those periods gradually. If either of the cats seems fearful—for example, crouching close to the floor—allow them to retreat to a place where they feel safe, and try again another day. If either cat growls, hisses or spits or shows any other aggressive behavior, shut the door to block visual contact immediately, preventing this from becoming a habit. Do not touch either cat at this time as they may redirect their aggression to you, simply because you are the nearest thing to them. Try this stage again on another day when both cats appear calmer and in a more positive mood.

For those cats coping well with hearing, smelling and seeing one another, the next stage is to allow them to interact physically. The starting point should be the same setup as you used when the cats met initially through the screen; for example, feed or play with both cats well away from the doorway that separates them. Then gently open the screen. Do not make a big deal about this—act calmly and quietly as if it is the most normal thing to do, even if you are secretly worried. Playing with each cat with a wand toy after they have eaten or grooming or stroking them, if this is what they enjoy, will reward each cat for being in the presence of the other and teach them that positive things occur when they see one another. You can systematically increase the time between delivery of such rewards, although do not forget to monitor each cat's arousal levels throughout. If these appear to be increasing, this is a clear sign that you have done enough for the time being: immediately close the connecting door. Next time, keep the session slightly shorter so that you prevent their arousal from increasing again.

If the cats choose to approach one another during one of these physical introduction sessions, allow them. If either cat appears not to be coping, lure each cat back to its own part of the house with food treats or a wand toy and close both the screen and the door between the cats. You can also utilize a target stick and ask the cat that seems to be coping least well to touch it with nose or paw to give his mind something else to focus on. Then, use the stick to move him away from the other cat without having to pick him up and physically restrict him. Make sure that the other cat does not follow (see Key Skill No. 3 for more details on how to lure and use a target stick). If your cat will engage in some target training in the presence of the other cat, it is a clear signal that he is not worried by the other cat.

Allow these introductions, which should always be supervised, to occur daily. Over time, the period the cats are together can be increased until the screen door can be removed completely. At this point, you can feed treats to both cats when they are close to one

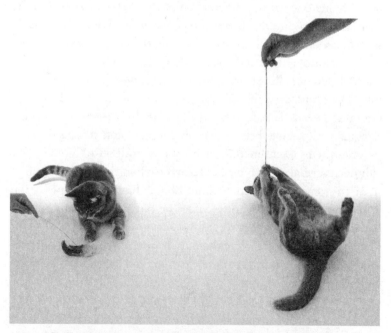

Playing with Cosmos and Herbie separately but side by side.

another or play with both cats at the same time. However, always ensure that each cat has his own set of resources in his own designated space, as this will greatly help them to perceive each other as no threat to their "possessions." It does not matter if the cats choose never to come in physical contact with one another: this should always be their choice and must never be forced. What is important is that their behavior is calm and relaxed in each other's presence.

MANY CATS THAT LIVE TOGETHER HAVE A ROCKY RELATIONSHIP, probably because they were not introduced to one another carefully enough or because a significant event caused a breakdown in the relationship—for example, when one cat spent a prolonged period of time away from the home due to a stay in the veterinary hospital. For most cats, however, it's possible to adapt the process for introducing a new cat. However, in this reintroduction process, instead of confining one cat to a single room, as if it was a cat new to the household, carve up your home into two distinct and roughly equally sized areas, one for each cat. For example, this may mean separating the house into upstairs and downstairs. If the cats have access to the outdoors, do not allow both outside at the same time during this reintroduction process. Likewise, if your cats previously had to share resources such as beds, food bowls and scratching posts, divide these as equally as possible between them: you may find you need to obtain new items to ensure that each cat has a full set to himself. You will need several of each type of resource when the cats do come to share the same space again, to diminish any lingering feelings of the other cat being a threat to resources by having a plentiful supply. Separating the cats will prevent any escalation of hostility between the cats, while giving each of them their own resources and physical space will allow both cats to develop their own individual territories without feeling under threat. It is a good idea to leave the cats separated like this for several days (and in some extreme cases, weeks) to make sure they are fully relaxed before any reintroduction occurs.

Now that each cat feels safer and more secure with his physical surroundings, you can start the reintroduction process following the instructions for introducing a new cat. Use the scent collection method (see Key Skill no. 7), as described earlier in this chapter, but instead of exchanging scents, first rub each cat's scented items onto his own individual belongings. For the previously shared items, this will cause them to smell predominantly of the cat who now has sole access to them; for any new items, this will reduce unfamiliar odors, increase the likelihood that the cats view them positively and greatly enhance both cats' feelings of security.

During the reintroduction process, try to maintain each cat's individual routine as much as possible. For example, if one cat tends to spend part of the evening on your lap in front of the television, maintain this routine by including this room in that cat's new quarters. You may find yourself split between two parts of your house for a period of time, and you may feel as if your routine has been disrupted unnecessarily. This will eventually correct itself, so persist: the reward of feline harmony throughout the house will be worth it! By teaching each cat that he no longer has to share his resources and that the presence of the other cat is a positive, not a negative, through gently desensitizing the cat to all the sensory components of the other cat and replacing them with positive feelings, feline harmony can be restored, leading to a happier household all around. In fact, this process can also be used to teach your cat that other animals, even those they may naturally be cautious of, such as a dog, can be friends, not enemies. For whatever reason, some cats seem very resistant to living in harmony with other cats or appear incompatible with particular cats, even after a reintroduction process: in such cases, it may be necessary either to find the most intractable individual a new home or divide the house up between the cats permanently (as in the first stage of the reintroduction process).

Few cats will never encounter another cat during their lifetime, and not all encounters are friendly. However, as we have seen in this chapter, making sure each cat feels familiar and secure with its physical surroundings is the best starting place for introducing two cats to one another. From there, gradual exposure to the other cat, one sense at a time where possible, and building an association between the other cat's presence and pleasurable experiences will allow the cats to learn about one another at a speed that they can cope with. Hopefully, they will learn not just to tolerate one another but will actually come to like one another. Such a technique can also be used with other animals, as we shall see in the next chapter.

CHAPTER 6

Cats and Other Pets

JUST AS CATS TEND TO RESORT TO THEIR WILD INSTINCTS WHEN encountering another cat, the same default is utilized when they meet an animal that they don't know much about. If the animal is small and looks like a potential meal, then most cats will switch to hunting mode, even if somewhat half-heartedly. Conversely, it must not be forgotten that although cats can be fearsome predators of mice and small birds, before they became fully domesticated they were also prey for larger predators, such as wolves or, more recently, stray dogs: hence most cats' instinctive distrust of dogs. In the United States, about a quarter of pet-owning households have one or two cats sharing their space with a dog. However, just because the home is shared does not mean the sharing is done amicably. Because dog-training classes and dog trainers offering one-to-one training sessions are readily accessible, there is always help on hand for owners wishing to teach their dog not to chase, stalk or herd their pet cats (not that all owners will have taken advantage of this). However, few owners will have considered whether their cat(s) can also be trained not to fear the dog. But of course they can![1]

Nowadays most pet cats have to contend with only one species that may wish to do them harm, and that is the dog: long gone are the days when cats, and especially kittens, were routinely at risk from foxes, wild dogs, big cats and other large carnivores. Nevertheless, there is probably some truth behind the old saying "fight like cats and dogs." Although most dogs are larger than cats, cats are, if anything, better armored, having sharp claws as well as sharp teeth. It should therefore pay dogs to avoid cats, and vice versa, but in fact cats tend to run away from dogs, and they usually fight back only when cornered. The antipathy between the two animals, although it probably has its origins in protective behavior that evolved prior to domestication, may have been further refined by both species' need to live close to people. Historically, when cats were allowed to breed willy-nilly and the loss of kittens was of very little public concern, unsupervised dogs were probably one of the cats' main enemies, potentially responsible for the death of many litters. If so, this would have reinforced the cat's tendency to defend itself against dogs and even to go on the offensive if there were vulnerable young nearby.

Luckily, another effect of domestication gives us the means whereby to temper, even reverse, this instinctive reaction. Just as young kittens are able to become socialized to both humans and other cats, principally during their second month of life, so they also have the capacity to become socialized to dogs—or at least to dogs that behave in a friendly way toward them. It's unclear whether kittens regard dogs as four-legged humans, as large clumsy cats, or as a separate category altogether. They may simply consider them to be similar to those cats that they get along with: Mike Tomkies, in his book *My Wilderness Wildcats*, recounts how tolerant his undomesticated Scottish wildcat kittens became toward his German shepherd dog Moobli, to the extent that they occasionally rubbed noses with him. Pet cats that cohabit with dogs routinely display the same affectionate behavior—tail-raising, head-rubbing, licking each other and sleeping together—behaviors that would normally be reserved for the cats in their own family group. Nevertheless, whatever cats conceive dogs to be, the individual encounters that they have with dogs during kittenhood seem to have a profound effect on their subsequent behavior toward other (similar) dogs.[2]

Because the kitten is very much the weaker and more vulnerable of the two, it is the dog's behavior that is usually crucial in determining the degree of trust that may develop. Because most kittens' natural reaction will be to flee or defend themselves from any unexpected advance, the introduction needs to be managed carefully, especially if the dog has had prior experience of—and probably enjoyed—chasing adult cats. Even more care and planning will be needed if the kittens' mother is herself not well disposed toward that particular dog, or dogs in general. Nonetheless, the kittens' future well-being will be enhanced if they are at least given the opportunity to learn a strategy for interacting with dogs other than simply running away or reacting with aggression.

Regardless of whether you already have a cat and are thinking of introducing a new dog, or vice versa, prudent preparation will greatly improve the chances of a successful relationship developing between a cat and a dog. Greyhounds and other dogs that have been bred to chase, and have subsequently had the opportunity to practice chasing small animals (or have even been trained to do so), are unlikely to make suitable companions for cats, as the motivation to chase is often too strong. Likewise, dogs that have been bred and worked as herding or stalking dogs, such as working sheep dogs or gundogs, may also be unsuitable: although they may not chase or try to catch the cat, most cats will find just being regularly stalked and stared at stressful.

For other breeds and types of dogs, it is more important to judge the dog's individual personality and usual behavior, rather than basing the decision purely on breed. There is often as much behavioral variation within a breed as there is among breeds. Thus, the chance of successful integration of a cat and a dog in the home is maximized by avoiding dogs that enjoy chasing, stalking and herding, or that show any aggressive behavior toward small animals. It is also a good idea to avoid dogs that are overly boisterous or playful (or at least to train them to control this behavior before introducing a cat). Likewise, it may not be fair, from the dog's perspective, to select a dog that is fearful of cats.

Similar considerations apply when taking the personality of the cat into account. A cat that is generally timid is not going to cope well with the addition of a dog. Equally, a confident cat that already shows aggressive behavior toward dogs is going to make any dog's life a misery. Ideally, a cat that is going to live with a dog should be confident and laid back, one that does not take flight easily, thus avoiding the instigation of a chase. The ideal personality for a dog that is to live with a cat is one that is calm, quiet and also laid back. In addition to personality, there are a number of other factors known to influence the likelihood of cats and dogs getting along. Positive experiences with the opposite species during kitten- and puppyhood will help develop a tolerant animal. If either animal has been chased or attacked by the other species, either during this early socialization phase or later in life, they are likely to have generalized their fear of that particular individual to the entire species, particularly if they have not had subsequent positive experiences with other individuals to balance this out. Therefore, it is important to glean whatever you can about the previous history of the animal you intend to introduce to your home, as well as taking into account the relevant situations that your current animal has experienced.

Personality and experience are important, but they are not the only factors worthy of consideration. Whether the cat and dog to be introduced are male or female appears to have no bearing on the relationship formed between them. However, the age of the animals does have an influence—introductions are generally more successful if the cat is less than six months of age and the dog less than a year: generally speaking, the younger the animal, the greater the chance of success. If you currently own a kitten or young cat and think you may get a dog during its lifetime, it is a good idea to allow visitors to bring calm, gentle dogs into the home. This way, your cat will learn early on in its life that dogs are not threatening, thus preparing it for the time when you make the decision to add a dog permanently to your household.[3]

INTRODUCING A DOG TO A HOME WITH A CAT OFTEN HAS A MORE positive outcome than vice versa—this is likely due to the cat already having established its territory within the home and therefore feeling confident enough to cope with the intrusion. A cat coming to any new home has to carve out his own territory and learn where his resources are and where the safest places are. Add a dog into the mix, and the cat's ability to learn that the new home is his territory and is safe may take considerably longer and result in some unwarranted stress, not only for the cat but also for the dog, not to mention their owner.

Prior to the new dog or new cat joining the household, it is worth making some adaptions to the home that will increase the cat's security (both real and perceived), thereby teaching him how he can keep himself safe. Initially, give your cat and dog separate living quarters within the home, using a physical barrier such as a door. If your cat has or will have outdoor access via a cat flap, it is wise to give him the exclusive use of the room it's in, to prevent any potential ambushes around the flap. If you do not currently have a cat flap but are thinking of installing one, it may be a good idea to install it into a window rather than at floor level, to give your cat a greater feeling of security while entering and exiting the home. If it is not possible to fit a cat flap into your home but you would like your cat to go outdoors, consider creating some sort of raised area on either side of the door where your cat will exit and enter the home. If the raised area is outdoors, your cat can survey the indoors (when the door is opened on his return), giving him the opportunity to check that his entry into the home will be safe before having to come down to dog level. The same holds true for when the cat is indoors and asks to go out—being able to sit on a platform (shelf, post or otherwise) that the dog cannot access will allow your cat to safely ask to go out. A cat that no longer feels safe when asking to go outdoors by his usual method of sitting in front of the door and meowing is likely not to have changed its mind about the outdoors, but simply does not feel comfortable sitting on the floor in case of ambush or other unwanted canine attention. Additionally, make sure that your cat has exclusive access to the places where his important resources are, namely his litter box, sleeping areas, food, water, toys and scratching posts.

If it is the cat that is new to the household, create a room just for him, as you would when introducing a cat to a house where one or more cats already live. If the layout of your home does not allow this, try to create an area that the dog cannot get to—this may be done by rearranging the furniture so as to block the dog's access to a certain part of the room. This should, as much as possible, include blocking the dog's ability to see the cat when it is in this area. Create a gap in the furniture just wide enough for the cat—but not the dog—to slip through, or provide access from above—using shelving or climbing posts. In addition, the area should have many hiding places to help the cat get out of view of the dog—sticking cardboard boxes together to make a larger hiding structure is an inexpensive way of providing additional security for the cat. Further alternatives include the use of room dividers or even creating a very large pen like those advertised as puppy pens—which is ironic, because such pens are designed to keep dogs in, not out! If using the latter, cover the pen's sides with fabric or cardboard to prevent dog and cat seeing one another. Over time, the new cat's access to more of the house can be widened as he becomes comfortable with his new routine and surroundings.

If it's the dog that's the newcomer, you may need to move some of the cat's resources in order to create distinct cat-only and dog-only areas. Cats often find alterations to their environment difficult, even changes that you know are designed to benefit their lives in the long run. Thus, make any changes gradually, and be sure that they are completed before the new dog arrives. It is also a good idea to think ahead to the point when your cat and dog will be sharing your entire home. If you provide your cat with places he can get to that the dog cannot, your cat will discover that he always has a means of getting away from the dog if he so wishes, thereby enhancing his perceived security and ultimately his happiness. The easiest way to do this is to provide plenty of hidey-holes, cozy places and vantage points in raised areas out of the dog's reach. Examples include cat trees with perches, cardboard boxes with cat-sized entry holes, and bedding placed on raised furniture such as tabletops and shelves. Even making sure chairs are not tucked under tables so your cat can use them to jump up onto even higher furniture and out of your dog's reach will help your cat feel safe.

It is important that your cat never feels he has to "run the gauntlet" to get to somewhere safe. If he feels vulnerable moving from one place of safety to another, then he is likely to do one of two things: he will either freeze in the dog's presence, or he will run from one safe place to another, thereby inviting the dog to give chase. Thus, adding secure "byways" throughout the house will help your cat to navigate through his territory while feeling safe. These can be achieved by having enough safe places dotted throughout the house that the cat never has far to travel from one to the other or even by constructing aerial walkways such that the cat can move from one part of the house to another out of the dog's reach. For the less adventurous, making simple changes such as leaving only a cat-width's space between the wall and the sofa will give your cat a dog-free byway that can then end in a piece of furniture, such as a side table, onto which your cat can jump, out of your dog's reach—that is, so long as your dog is wider than your cat.

Some cats living alongside dogs will always require an area that is their own so that they do not feel that their resources are under threat. Threats can occur in the form of the dog stealing the cat food, disturbance while using the litter box and (with small dogs in particular) bed-napping. Although resources can be made dog-proof by having them at heights dogs cannot reach, for some cats this is not feasible. For example, elderly cats may find jumping up onto raised surfaces difficult, and in some homes there is not enough space to make all resources raised. Excluding the dog from one room or area in your house is one way of overcoming these potential problems. Fitting a cat flap in a normally closed door to a cat-only room is an ideal way to create a dog-free zone that gives your cat complete privacy to eat or go to the litter box. Remember, however, that the cat's food bowl and litter box should be as far apart as possible within this room. For dogs the same size as or smaller than the cat, the cat flap may need to be controlled by microchip or magnetic collar to prevent canine access.

Partial barriers can be useful for introducing dog to cat, and vice versa, in a controlled manner, complementing the use of more permanent divisions of the house. A baby gate is ideal for this—as long as it is high enough to prevent your dog from jumping over it. Some are made with an integrated cat flap that allows the cat to pass through without your dog being able to follow. Make sure, however, that the cat flap is not of a size that your dog could get his head stuck in. If you lack a suitable doorway in which to place a baby gate, you can instead divide the space from wall to wall—one way is to create a "wall" of furniture, leaving a gap through which the dog and cat can view one another. This space must, however, be blocked by something that prevents physical access to one another but allows the animals visual contact—one side of a puppy pen or baby playpen can work well.

Whether the dog is already the resident or is about to be introduced into the household, it is really important that he has already learned several basic skills. These are to perform the following behaviors when asked: "sit," "settle," "leave," "give me your attention" and "be quiet." In addition, it will be greatly helpful if the dog is already crate trained. Having a dog that feels safe and secure in a large crate means, if there is no possibility of separating two rooms by a physical barrier (for example, your home is completely open plan), that a reasonable alternative is to allow the cat free access to the home at the first introduction while the dog is securely shut in his crate. As an alternative to—or in addition to—a crate, ensuring the dog is comfortable wearing a harness to which a house line can be attached when he is in the same room as the cat without any physical barrier will provide extra security for both animals during early introductions. A house line is like a lead but is lighter and longer, allowing the dog to drag it behind him without disruption to his behavior, and allowing you to move the dog safely if needed. You need to make sure that you can keep your dog quiet and calm in the presence of the cat and, when needed, that you can keep his focus on you and, in case he picks up any of the cat's "belongings," can drop them on command.[4]

Before introducing a cat to anything new, whether that be another animal or a physical situation, it is always important to stop and consider whether the new experience could worry the cat in any way. If so, all the different sensory properties of the stimuli should be examined and thought given to how on first introductions these could be diluted or teased apart in any form and associated with positive consequences—the foundations of Key Skill No. 2, systematic desensitization and counterconditioning. Furthermore, the dilute forms of the stimulus should be introduced in a manner where the cat always feels in control—this involves allowing the cat to observe, approach and explore the stimulus at his own pace in his own time—the foundation of Key Skill No. 1.

Once cat and dog are both settled into a routine in their own parts of the house, and prior to them catching sight of one another, the first step in the introduction process is to allow them to learn each other's odor. This is done by exchanging items that smell of the other. As for cat-cat introductions, cotton gloves, grooming brushes, or a clean cloth placed in the animal's bedding can be used to collect the scent from the cat and from the dog (see Key Skill No. 7 for more details). Choose a method of scent collection that best suits each individual so that it is a positive experience for both animals.

At a time when your cat is relaxed, place the dog-scented item into your cat's quarters and watch his reaction. The ideal response is that your cat sniffs it a couple of times and takes no further interest. If, however, your cat appears on edge or startled or actively avoids the scented item—or worse, appears hostile toward it—remove it immediately and wait a few days before trying again. Try again with the same scented item, as it is likely to have lost some of its concentration of smell over those few days and your cat may now be more likely to accept it. Once your cat is showing no obvious interest in the scented item, you can increase the concentration of scent by, for example, rubbing the item on the dog for longer or leaving it in his bed for longer periods of time. When your cat appears fully relaxed with this scent in his general environment, you can progress to placing the scented item in his bed to allow his scent to also gather on the item. This will help mix his scent with that of the dog and teach him that the dog is part of his social group, because the dog's scent can now be found alongside his own. The same procedure should also be carried out exactly in reverse to introduce your cat's scent to your dog.

First impressions are crucial—if either the dog or cat is spooked during their first actual physical encounter, later meetings are less likely to go well. Thus, in order to reduce the chances of a negative first encounter, the initial introductions should take place at the boundary of the cat-only and dog-only areas, allowing the cat to stay within the area of the home he feels most safe without the dog intruding into this space. At this stage the animals should be able to see one another but have no opportunity to interact physically. If the door between the animals' living quarters does not have a glass panel within it, then consider inserting a screen in the doorway or utilizing a baby gate covered in mesh (to allow the animals to see one another but not be able to physically interact). Provide the cat with safe places near this boundary, ideally raised, from which he can survey the situation. A cardboard box or igloo-shaped bed raised off the ground will help the cat to feel safe while at the same time enable him to observe the dog and gain confidence in his presence. If the cat is a kitten, a floor-level hiding place with a small entry hole may be easier for the kitten to access quickly, rather than having to climb off the ground. Cats generally feel safe if they feel they are hidden, and in order to feel hidden, they need at the very least their line of sight blocked from the thing they may feel afraid of. Ideally, the cat should have a safe place from which he can take fleeting glances at the dog as he chooses.

It is also a good plan to provide a safe place for your dog, such as a crate with a blanket placed over the top and down the sides to which he can retreat if he finds the cat in any way threatening. This should be your dog's safe place. A helper really is invaluable at this stage, as you will need someone with the dog to manage his behavior during these visual introductions, as well as another person on the same side of the barrier as the cat to provide rewards at precisely the right moments. In order to maximize the chances of your dog behaving in a calm manner, it is a good idea to have taken him for some exercise prior to the visual introduction to your cat.

Your dog handler should engage the dog in a calm and positive task, so that his focus is anywhere but on the cat: this could be something like eating from a puzzle feeder, simple target training (where the dog is taught to touch its paw or nose to a target stick or handler's hand) or simply lying down and relaxing on command. Avoid the use of any activities that may be rewarding for your dog but your cat could perceive as frightening, such as playing with squeaky toys. Your dog should initially be wearing his harness and house line: the house line is there as a safety net in case any previously learned commands fail and the dog needs to be moved away from the doorway, screen or gate. As your confidence grows in your cat's (and dog's) ability to remain calm in each other's presence, you can remove the house line, although it can be a good idea to keep the harness on for a while in case you do need to restrain your dog quickly.

You will need to monitor the cat's reaction and respond appropriately. If your cat decides not to use his safe place or chooses to come out of it, that is an indication that he is relatively comfortable with the dog being in view. Reward him for such behavior using high-value food treats or play—play should ideally be directed out of visual sight of the dog in order to prevent frustration at being unable to join in the game or overexcitement at the sight of toys. Avoid any rewards that entail physical contact between you and the cat at this stage, just in case the dog startles the cat and he inadvertently redirects his startle as aggression toward you. You can, however, talk to your cat in a calm, gentle voice to reassure and praise him. If your cat is not comfortable coming out of his safe place, do not try to force him or lure him with rewards. We do not want to put him in a position where he suddenly feels uncomfortable and panics. Each decision to look at the dog, or move closer to him, should be his own and rewarded immediately (Key Skill No. 1).

Initial sessions should be short and frequent to keep both cat and dog arousal levels in the safe zone, but they can be gradually increased

Squidge is busy with a puzzle feeder
while Herbie watches, safely behind the barrier.

over time. As sessions get longer, giving your cat something positive
to do while in front of the dog will help build his confidence around
the dog. Additionally, his confidence will be given a further boost if
he finds that nothing bad happens when he takes his eyes off the dog,
reassuring him that he does not need to remain in a heightened state
of vigilance whenever in the visual presence of the dog. A puzzle
feeder is a useful way of achieving this: in order to gain food from the
feeder, the cat has to divert his attention away from the dog for a
moment. Thus, as he accesses food from the puzzle feeder, he will
automatically be rewarded for taking his eyes off the dog.

Building on these initial steps, we can now provide more positive experiences for both animals while they are in each other's presence. This will teach them that positive things occur when they are near each another (Key Skill No. 2). In practical terms, the next step is therefore to remove the physical barrier between cat and dog, although at this stage it is a good idea to keep your dog on the harness and house line (which your handler holds) so your handler has some control of where he goes. Keep your dog in the part of the house that you have dedicated to him, and position yourself in the same room as the dog but at a distance from him. Your presence there will help encourage your cat to come into that space in his own time. Again, ideally, you should have a handler to look after your dog's needs and training for this part of the introduction, so as to keep your dog's attention intermittently diverted from the cat.

Do not worry if your cat initially chooses not to come into the room—he will be learning that the barrier that used to separate him from the dog is no longer there, and he may need time to feel confident enough to get closer to the dog without the protection of the barrier. For some cats, it may take only minutes before they are in the room with you; for others, it may take several repetitions of removing the barrier over a period of days or even weeks. Just be prepared to go at whatever speed your cat chooses. Picking him up and bringing him in the room will only remove his ability to decide for himself, potentially making him feel uncomfortable and thus undoing all the good associations he has previously learned about the dog. When he does decide to venture into the room, you can again reward this behavior with a food treat or game. At this point, your cat may become quite tentative again, so it is really important that your dog handler keeps your dog's attention anywhere but on the cat.

Some cats are really inquisitive and will want to get close to the dog, simply so they can investigate him. However, such boldness can be risky—the cat may get itself into a situation where he has approached the dog to investigate him, then realized that he is not that comfortable being so close to the dog and panics. The cat may then lash out at the dog or simply turn and flee at great speed. The dog may then be provoked into chasing the cat, which is precisely what the whole process has been designed to avoid. Having the dog on a house line means you can monitor both the cat's and the dog's behavior carefully, and pre-emptively move the dog away if the cat is getting too close. In this way you can create more distance between the two animals, thus reducing the probability that the cat will overreact. Having the cat free to move increases his perception of safety. Cats do not like to feel physically restricted: dogs are much better at coping with a harness and lead. Having the cat loose also allows him to jump up high or enter a safe place such as a conveniently placed cardboard box if he desires. Some cats may keep moving closer to the dog as you move the dog away: in such a situation, you can wave a wand toy quickly along the floor, luring the cat away from the dog.

It's important never to let your cat enter your dog's crate or bed, as this is the dog's safe place. If you see this situation about to arise, again, lure your cat away with a wand toy. If you have a dog that wishes to play with one of the cat's toys, make sure his handler engages him in something he finds equally as rewarding—a game of his own, a chew treat or some fun training.

Avoid activities that get the dog too excited, active or noisy, as any of these may upset the cat. Calming activities for your dog can include grooming or stroking, or if you can't give him your full attention, you can provide a puzzle feeder filled with meat paste that has to be licked out. However, if your dog is possessive around food in any way, it is best to avoid chews and puzzle feeders in case the cat's presence makes him feel threatened.

The key to success is that both cat and dog should always feel comfortable in one another's presence, each doing his own thing. The goal is not necessarily to have them physically interacting—if they choose to do so in a positive manner, this is an added bonus, but both parties must voluntarily participate in such an interaction. Cats and dogs can get along well enough to choose to sleep together and groom one another: however, such behavior usually occurs only if the two have been brought up together and have lived together for a considerable time. In most instances, what we can aim for is to have the cat and the dog in the same room, both comfortable and calm in each other's presence, both going about their everyday lives and sharing the same space.

Thus, while the dog is on his house line with a handler and engaged in a quiet activity, try inviting your cat to take part in an activity he enjoys—for example, invite him onto your lap for some fuss, give him a puzzle feeder (at a distance from the dog), groom him or play with him. If your cat is not keen to engage in any of these

Cosmos is busy with his homemade puzzle feeder
while I reward Squidge for remaining calm.

activities, just let him be; it may be that he is comfortable enough to occupy himself in the dog's presence or that at this stage he would prefer to keep a watchful eye on the dog, and as his confidence increases, he will engage in other tasks.

Until you have given your cat many opportunities to be in the same room as the dog, it is advisable to keep the animals in their own parts of the house when you are not able to supervise. Once your dog and cat have been in the same room many times, with both seeming calm and relaxed, you can let the dog investigate the cat's part of the home. Do this first of all with the dog on the house line, and if all is well, subsequent investigations can be unrestricted (though still supervised). If both dog and cat appear calm and relaxed, you can leave the barrier separating the two areas of the home open. If you decide to have a permanent cat-only room, this should always be out of bounds for the dog. Even if cat and dog become the best of friends, do not be tempted to remove some of the cat's safe places and hidey-holes—these may be the foundations of such a good friendship. As dog and cat integrate, you may wish to spread both the cat's and the dog's belongings throughout the house—this should be no problem for them as long as it is done gradually and several belonging to the cat are kept in places the dog cannot reach.

BECAUSE DOMESTICATION OF THE DOG HAS BEEN A LONGER AND more complete process than that of the cat, the dog has lost some aspects of its hunting instincts. This is not, however, the case for the cat. Small "furries" such as hamsters and gerbils, caged birds and rabbits are never going to be the best of friends for cats. They are ultimately too similar to their wild counterparts, which cats are instinctively motivated to catch. In turn, these types of pets are instinctively afraid of cats, just as cats are naturally interested in them. For some cats, particularly those that are successful hunters, small caged animals have instant appeal, and given the chance, the cat would chase and pounce on the loose animal. Other cats, while appearing mesmerized by birds fluttering in their cage or a gerbil scurrying around in its tank, appear not to know what to do. Then there are those exceptional cats that show no interest whatsoever, but even these can cause stress to the small "furries" or caged birds, which instinctively recognize cats as predators (by their appearance and also probably by their odor). The cat can also become stressed through feelings of frustration—a gerbil that your cat can watch all day every day but can never catch may infuriate the cat.

Training the cat not to view the family's pet mouse or hamster as a tasty snack is a more difficult task than training a dog to leave the cat alone. In the cat, the switch into predator mode seems to be triggered by a small number of visual features, many of which are emulated in commercially available "toys" for cats. These include movement (fast and in straight lines is best), size (bigger than a fly, smaller than a rat), presence of limbs (apparently the more the merrier—some cats love to play with faux spiders), short bursts of high-pitched sound and texture—fur or feathers. (Given the cat's acute sense of smell, odor probably also plays a part, but this is not well understood.) Any one of these features can be powerful enough to trigger cats into chasing and pouncing, especially if they're young and inexperienced, as when kittens chase fallen leaves blown past them by the wind.

Cats' seemingly irrepressible predatory instincts mean that they are unlikely to make reliable companions for small furry pets, such as hamsters and gerbils, or caged birds. Occasionally an account will surface of a mother cat whose maternal instincts have been hijacked by the young of what would normally be prey species, such as a baby squirrel. However, such instances are very much the exception. It may be possible to train a cat to ignore any small animals that share his owner's house, but because just one lapse might literally be fatal, permanent physical separation will usually be the safest option.

Training can go only so far to remedy this incompatibility. The best we are likely to be able to do is to teach the cat that there are other things in the house that are more interesting than the bird cage or gerbil tank, thus diverting his attention elsewhere. We cannot "turn off" the cat's predatory instinct, and for this reason cats should never be left unsupervised with such small pets.

There are, however, some things you can do to persuade your cat that your pet mouse or bird is not that interesting. If your cat shows interest in the animal's housing, try to block your cat's view by placing a curtain or barrier around it every time your cat shows interest in it. The ideal situation for your small pet is to place its housing where your cat does not have access. You can also teach your cat an alternative behavior to looking at the cage, such as looking at you (as described in Chapter 5). Make a short, sharp squeaking sound by pursing your lips and sucking air through them. This may be an inviting sound that you already use with your cat. If so, he is likely to turn round and look at you. If not, the novelty of the sound is also likely to encourage him to turn toward you to see where the noise came from. Once he does look at you, mark this behavior (see Key Skill No. 4) so your cat knows what behavior is the right behavior, and immediately reward with a piece of food. Repeat this many times, spaced over several training sessions, until you are confident that your cat will always look at you when you make the sound. At this stage you do not need to reward every look at you with food or whatever reward you have chosen, thereby helping to maintain the response over the longer term (see Key Skill No. 8).

Having this tool at your disposal allows you to gain your cat's attention whenever you need to: in this instance you can divert his attention away from the cage and toward you by, for example, showing him a wand toy you already have in your hand, once he has directed his gaze to you. However, there will be situations where a toy is not to hand and you need to get your cat away from the cage. Teaching your cat to come when called (see Chapter 10) will give you the tools to be able to move your cat away from the cage without having to go and pick him up. Remember, your reward has to be greater in value than the reward the cat gains from watching your small furry pet or bird, or he will not be motivated enough to move away from the cage. As well as having already learned how to divert your cat's attention and move him away from the cage, it may be necessary to fulfill his interest in small furry or feathered animals in other ways. When playing with your cat, include lots of predatory type games (wand toys; chasing balls; opportunities to capture, bite and chew toys made with fur and feather; puzzle feeders). This provides an outlet for predatory behavior that should satisfy your cat's needs in a welfare-friendly manner, that is, where no animals are harmed or stressed. It will also provide an outlet for mental and physical energy, thus leaving your cat more likely to want to sleep than watch the cage in his spare time.

WHETHER THE FELINE INSTINCT IS TO HUNT IT OR FLEE FROM IT, cats can be trained to live alongside other animals as long as alternative appropriate outlets for hunting instincts are provided and the cat's perceived security within the home is always optimal. Dogs have long been described as man's best friend, but with systematic desensitization and counterconditioning and allowing the cat to always be in control of how close to get to the dog, the cat may come to rival the owner as the dog's best friend—it is not unheard of for dogs and cats that have been appropriately introduced and taught to enjoy each other's company to curl up together in the same bed. Some cats even accompany their family dogs on their walks. The training techniques of systematic desensitization and counterconditioning are not confined to learning about living things—they are equally useful for training cats to cope with all sorts of objects they may encounter in their lives but do not necessarily feel immediately comfortable with. The cat carrier is one such example and is the topic of the next chapter.

CHAPTER 7

Cooped-up Cats

I T'S A RARE CAT THAT CAN IGNORE A CARDBOARD BOX. AN EMPTY
one, of course. Once the owner has removed the contents and has
turned her back to put them away, somehow the cat materializes
as if from nowhere, peering out of the box. On YouTube, millions of
cat lovers have watched Maru, a Scottish Fold from Japan, trying to
cram himself into smaller and smaller cardboard boxes: feline star
of Twitter, @MYSADCAT (a.k.a. the Bear), regularly appears in
boxes decorated with captions such as "Hotel Catifornia."

Yet what greater contrast in behavior could there be when a
superficially similar enclosure—the cat carrier—is brought out? No
cat to be seen, or just a brief glimpse of his tail disappearing out of
the room. This poses a problem for the cat owner—cats often need
to be transported—for example, they may need a visit to the vet,
or they may be booked into a boarding cattery while their owners go
on vacation. Either way, transporting the cat in a cat carrier—
regardless of whether travel is by car, train, foot or even plane—is
the safest and often easiest means of getting a cat from A to B. However, if the cat won't even entertain the idea of going into the carrier, all your travel plans will be thwarted.

So why is it that for so many cats, cartons are irresistible but cat
carriers are abominable? On the surface there is little to differentiate
them (they both provide a confined space), so the cat's reaction
must somehow stem from his experience of both.

To begin with the cardboard box, we stumbled upon the explanation by accident during an apparently unrelated study. John's team was examining what kinds of surfaces cats in rehoming facilities like resting on, and to test the effects of rain on outdoor pen furniture, we constructed open cubes with the test material—carpet, for example—on both the upper and lower surfaces. And, yes, we did confirm that cats don't like sitting on wet carpet! (They don't mind damp wood.) However, whether they settled on the dry surface inside the cube depended not at all on what that surface was but on which way the cube was facing. The cats liked resting in a partially enclosed space, but only if it allowed them to keep an eye on the other cats, mainly those in the same pen or the pen next door. Many other studies have followed this initial discovery, showing that such boxes significantly reduce stress in rescued cats, culminating in widely used products such as the British Columbia Society for the Prevention of Cruelty to Animals' Hide, Perch & Go box and Cats Protection's Feline Fort.[1]

Why should pet cats find such fascination in enclosed spaces? It's perhaps understandable that a cat that's recently been uprooted from its accustomed territory might wish to hide away until it has made sense of its new surroundings, but it's less obvious why a cat that is in its own familiar home would want to do so. The likely answer can be traced to two otherwise unrelated factors: the cat's small size and its sharp claws. Cats, although predators in their own right, also have enemies. Before they were domesticated, those might have been wolves or bears. Domestication largely removed cats from such threats but introduced new ones, including other, more aggressive cats, stray dogs and even cat-phobic humans (not to mention the imaginary threats posed by inexplicable events such as firework displays). Before cats can sleep soundly, they need to feel safe, and so any feature of their environment that might offer safety is worth investigating, whether it's a corner of their owner's duvet or a cardboard box. (For this reason, a cat that is frightened of loud noises—and most are—should be provided with a quiet place to hide come fireworks season.)

Unlike many other carnivores, cats are reluctant to dig their own shelter. Badgers, wolves, foxes and dogs have broad, blunt claws that provide traction when they're running and also allow them to exca-vate burrows for themselves. Cats, big and small (except the chee-tah), have sharp claws that are their primary hunting weapons, which when not required are sheathed to preserve their sharpness. This specialized hunting equipment therefore prevents cats from do-ing any serious digging. In the outdoors, cats have to seek out natu-ral crevices or holes dug by other animals: the domestic cat's wild counterparts in North Africa favor the abandoned burrows of fen-nec foxes. Ready-made enclosed spaces of the right shape and size are few and far between in nature, so it pays cats to investigate any potential candidate at every opportunity. Even today, when cats are better protected than at any other time in their history, this instinct is evidently still strong.

So what is it about the cat carrier that turns cats into disappearing acts? They are designed to be the right size for a cat, and once inside the cat is protected on three sides and from above, just as it is in a cardboard box lying on its side. The difference, obviously, is that cats will have found that they can jump in and out of cardboard boxes whenever they like, whereas the cat carrier soon becomes associated in the cat's mind with all manner of unpleasant occurrences—the struggle with the owner prior to being forced inside, the door being closed (resulting in a sense of being trapped), the feeling of instabil-ity as the carrier is lifted and carried, the subsequent car journey to the veterinary clinic (a place smelling of dogs and frightened cats, and where unwelcome manipulations occur) and so on. For all we know, the carrier may still contain traces of the smell of the cat's fear (likely deposited from the cat's scent glands in the paws), however well it has been cleaned out since its last use. Compare these scenar-ios with those of a kitten who has only ever experienced an open cat carrier in the home as somewhere to sleep and play and is therefore unlikely to have developed any negative feelings toward the carrier.

The adverse reaction is therefore very much a product of learning. There's nothing intrinsically wrong with the carrier itself, so long as it's been well designed, but its connotations in the cat's mind quickly become overwhelmingly negative. The trick to preventing these associations from building up is to give every cat the opportunity to explore its carrier before it has ever been used for actual carrying. If the cat has already become frightened of the carrier, then that association will have to be broken. Training can be used to reinforce the idea that the carrier is fundamentally a safe and pleasant place to be in, gradually building toward the stage where the door can be closed, and subsequently to the point where the carrier is lifted off the ground, with cat happily ensconced within.

Helping your cat learn to love the carrier begins with ensuring that he finds his carrier safe and inviting. If your cat has already decided he very much dislikes his carrier, this could be down to several things. One is the physical features of the carrier: some are quite small, and cats may find being crammed into a small space uncomfortable and unsettling—your cat should have enough space in his carrier to stand up and turn around. Some carriers have a lid facing the ceiling, requiring the cat either to be lifted into the carrier or to voluntarily jump in—ideally, your cat should be able to walk freely into the carrier from the floor, so entry doors at ground level are best. Carriers also vary in the amount of visibility the cat has once he's inside. Some carriers are very open—for example, those made of plastic coated metal mesh—while others allow the cat to be partially concealed, providing limited visibility through a mesh door and open slits on plastic sides. The latter type of carrier will partly hide the cat and hence help him feel secure.

The odor of the carrier may be irrelevant to us, but it is of great significance to the cat. If a cat has already had unpleasant experiences in this particular carrier, such as a toileting accident, or associates his carrier with an unpleasant destination, for example, the vet's, it is likely that the carrier will have absorbed smells that will act as reminders of the unpleasant experience that the cat had. Such reminders may include the odors of the cat's scent glands—which, although we cannot smell them, are all too easily detected by the cat. People often say animals can smell fear, and this time they may well be right. Other reminders may include the remnant smell of the toileting accident, which despite your vigorous cleaning, cats can often still detect. Even the smell of the cleaning product itself can become aversive to the cat, although insignificant to our noses. Carriers made of wicker or fabric are most likely to become impregnated with such smells and are hardest to clean—thus they should be avoided, however cute they may seem.[2]

Paradoxically, once the cat is inside the carrier, he may be reluctant to come out again. In the past, you may have struggled to get your cat out of his carrier and onto the veterinary consultation table: the cat has apparently decided that being in the carrier, although not ideal, is actually safer than venturing out into the unknown expanse of the strange-smelling veterinary table where he can see an unfamiliar face peering back at him. As a result, the cat may have been pulled unceremoniously out of the carrier, or the carrier tipped in attempts to evacuate him. Such experiences only confirm to the cat that the carrier is a place to be avoided at all costs. We will work on teaching your cat to feel comfortable enough being in new situations that he will voluntarily exit the carrier. However, there may be situations in your cat's life where he will not want to leave the carrier under any circumstances—for example, during an emergency trip to the vet's following an injury. Not only would pulling the cat out or tipping the carrier be distressing for the cat, it may also cause him further pain or injury. A carrier in which the top part can be easily unclipped allows the cat to remain in the lower half, where he can be examined and removed only if needed.

THE PHYSICAL FORM OF THE CARRIER IS NOT THE ONLY PROBLEM, possibly not even the major problem. Several situations relating to the carrier can lead to negative associations being created in the cat's mind. First, if the only way you have been able to get your cat into (and possibly out of) the carrier has been by physical force, then your cat is likely to view the carrier negatively: he may then struggle even more on each subsequent attempt to get him into the carrier. Therefore you should eliminate any behavior that involves force to get your cat into (or out of) the cat carrier. Also, it is wise to start training well in advance of any planned uses of the carrier, such as, for example, scheduled veterinary visits or a stay in a boarding cattery.

Second, if you have more than one cat, do you ever use the same carrier for your different cats? This could involve more than one cat traveling at the same time in the same carrier, or each cat traveling in the carrier at different times. Although cats may seem to get along very well in the home where resources and space are ample, forcing them to share a small space, and one that they may already be frightened of, is a recipe for disaster. Likewise, if different cats share the same carrier but use it at different times, the scent left by one frightened cat will be detected by the other, and may signal fear to that cat.

Finally, the purpose of the journey is likely to impact on how negative or positive your cat finds the carrier. If the only time it is used is to transport the cat somewhere he finds unpleasant (the vet's, the boarding cattery, the groomer), then he will quickly learn that the carrier predicts that such a trip is imminent. Recall that in operant conditioning a cat who discovers that his behavior leads to a negative consequence is much less likely to perform this behavior again—in this case the behavior of entering the carrier conjures up memories of a negatively perceived destination. Thus, your cat will start to feel anxious at the mere sight of the carrier and avoid entering it at all costs, even if the next destination is to somewhere positive (of course, your cat cannot know this). We can begin to break down such associations by keeping the carrier routinely accessible to the cat in the home and pairing it with positive things such as treats and toys (using Key Skill No. 2, systematic desensitization and counterconditioning).

Likewise, traveling in the carrier using motor transport can be very stressful for cats—the noise of the engine, the strange vibrating movement and the glimpses of the outside world flashing past at high speed may be deeply unsettling. Without careful exposure to each of these individual elements of travel, initially at very low intensity and accompanied by high-value rewards (another use of Key Skill No. 2), cats can quickly become terrified of traveling and associate the cat carrier with such travel, and subsequently they will refuse to go near the carrier, even in the home.

Taking into account all the possibilities for previous negative experiences, training is likely to proceed more smoothly if you buy a new carrier, even if your current carrier fits the criteria already mentioned. The new carrier will not have such strong negative associations attached to it, thus making the task of reteaching your cat to like the carrier easier. For cats completely new to the carrier, getting the most inviting one will simply give you a head start.

Teaching a cat to enjoy being in the carrier, and then also traveling within it, cannot be done in a single step. The cat has to learn to associate a wide variety of sensory information with positive feelings, not negative ones. The key to success is to break down the tasks into small achievable goals (see diagram on next page). This way, the cat is never asked to learn too much at any one time, preventing him from becoming overwhelmed or reverting to fearing the carrier.

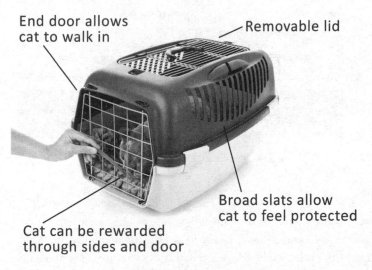

End door allows cat to walk in

Removable lid

Broad slats allow cat to feel protected

Cat can be rewarded through sides and door

The ideal cat carrier.

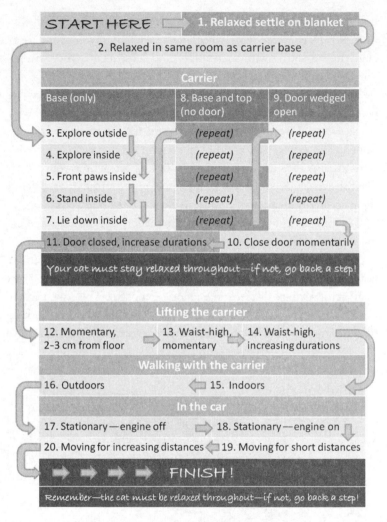

The twenty steps to traveling in a cat carrier.

The crux of mastering cat carrier training is teaching your cat to relax on cue—not only when he's inside it but also when he's outside. A cat is much more likely to cope with being confined inside the carrier if he is relaxed. Being in a cat carrier involves an element of behavioral restriction (the cat knows he is locked into a box), and in many cases this restriction lasts for prolonged periods of time—some people may be lucky enough to have their vet's around the corner, but most have to travel some distance. Furthermore, trips such as holidays or visiting friends (with the cat) or moving house may involve even greater distances. By ensuring that the cat is relaxed before even introducing the cat carrier, we are more likely to be able to teach the cat to remain relaxed around and in the cat carrier. Teaching a cat to relax (Key Skill No. 6) is described fully in Chapter 3 but in summary involves providing the cat with a comfortable surface on which to relax, such as a blanket (something that fits into the cat carrier), and rewarding successive approximations of relaxed behavior, thereby shaping the cat's behavior toward the double goal of being mentally and physically relaxed. Approximations of this goal may involve him standing on the blanket, sitting on it, circling on it (in preparation to lie down), lying on it and remaining in a restful and relaxed state while lying on it. Giving careful consideration to the cat's body language, facial expressions and vocalizations as well as posture before deciding whether and precisely when to reward will help you to reward the emotional state of relaxation, rather than simply just the physical location and posture the cat finds himself in. Relaxed cats may slowly blink their eyes or shut them completely; they may be silent or purr gently. Their tails will be held loosely from their body and if there is any movement, it will be slow and gentle. Their breathing will be slow so that the chest rises and falls gently, and when lying down, a very relaxed cat will turn his paws upward so that the pads are out of contact with the surface he is resting on.

When teaching a cat to relax, rewards should be selected carefully so as not to excite the cat too much. There is a fine line between offering rewards that are motivating enough for the cat to work for them versus making him too excited in anticipation of their delivery and thus not being able to relax. Thus, keeping within optimal engagement (calm but interested in you and the training task, described fully in Chapter 2) is particularly crucial when teaching a cat to relax—we are asking the cat to experience an emotional state (relaxed) at the same time as performing a behavior (lying down in the carrier). One way to prevent overarousal due to the anticipation of rewards such as food is to switch between several different rewards—for example, one that is calming but perhaps less motivating (stroking, for instance) can be alternated with one that is highly motivating but also a little more likely to increase excitation (for instance, food or toys).

Although it may seem like a lot of work to teach your cat to relax on a blanket when the real aim is to get him to accept the cat carrier, once you have gotten him relaxed, you will find that half of the hard work is already done. By placing the blanket in the cat carrier before introducing the carrier, your cat will be more likely to venture into it, as he already associates his blanket with a nice, relaxed, safe feeling.

If your cat has not actively avoided the cat carrier when he has seen it on previous occasions but likewise has never voluntarily gone inside it, we can teach him that it is a positive place to be by simply associating it with pleasant things.

The first step is to place it in a quiet, secure and comfortable part of your home—somewhere that your cat regularly spends time in, and not the place where you have previously tried to get him into the carrier—in case he has already made a connection among the location, the carrier and a lack of control. Remove the door and upper half of the carrier, and place the blanket you have associated with relaxation into the carrier, before your cat has a chance to see it. Impregnating the carrier with your cat's own scent (Key Skill No. 7) by rubbing a cloth on your cat's facial areas when you are stroking him (but only if he enjoys being stroked) and rubbing this cloth on the outer corners, entrance and inside of the cat carrier will also help your cat perceive the carrier to be a familiar and secure place—even though it will still look exactly the same to you. Place items your cat really values in, around and leading up to the carrier. For example, you could place a few of his favorite treats or a sprinkling of catnip up to and into the carrier. You can also place his food bowl with his meal in it inside the carrier so that by following the lure trail of treats or catnip, your cat finds himself in the carrier with an even bigger reward—his meal. This is an ideal scenario to set up for when you are not there, leaving the cat to voluntarily explore the carrier in his own time.

When you are there, you can provide the lure yourself; for example, you might drag a long feather or wand toy into the carrier and then play a game with your cat once he is in the carrier. For more information on how to perform luring, see Key Skill No. 3. Once the lure has enticed your cat inside the carrier and the reward of the game or meal has ended, the sight and feel under paw of the previously conditioned relaxation blanket in the carrier will encourage the cat to settle down into the carrier rather than immediately vacate it. Ultimately, we want to encourage prolonged visits to the carrier because they will be what the cat experiences when the carrier is taken out of the home.

CATS WHO HAVE HAD PREVIOUS NEGATIVE EXPERIENCES WITH A carrier, and cats that are naturally quite wary or anxious, are unlikely to be confident enough to explore the carrier voluntarily, even with the use of lures. Therefore, a different teaching method is needed for such cats, one that dilutes the impact of the cat carrier, thereby giving you a chance to create new positive associations in your cat's mind—by the process of systematic desensitization and counterconditioning (Key Skill No. 2). Ensuring the relaxation blanket is used throughout the process will help to keep the cat below the threshold of fearing the carrier, which you will be reducing further by initially exposing him only to the various components it is made from.

Begin your training with rewarding for relaxed behavior on the blanket, with the cat carrier placed some distance away. The distance between blanket and carrier should depend on your cat's current perception of the carrier: if it is very negative, make sure they are as far apart as is feasible—say, in opposite corners of the room. Gradually, and when your cat is not using it, move the blanket toward the carrier. (Never move the blanket while your cat is on it, as that will likely spook him.) Once your cat looks relaxed on the blanket, you can mark this with the word "good" and then provide your chosen reward (toy, food treat, fuss) off the blanket. The use of a verbal marker allows you to tell the cat the correct behavior that led to the reward while being able to provide this reward off the blanket. The importance of giving the cat the reward off the blanket at this stage is that it allows you to move the blanket to its next location while your cat is focused on its reward. A verbal marker becomes predictive that the real food reward is on its way, through the process of classical conditioning (see Key Skill No. 4, marking a behavior). Its use will buy you time to provide the reward while enabling you to be precise in the timing of its delivery. Always go at a pace your cat is comfortable with, tailoring the length of session to your cat's engagement and always ending on a positive note—for instance, with a good example of the behavior you desire (see Key Skill No. 9, how to finish any training task).

Over several training sessions, remove the top of the cat carrier and hide it somewhere nearby. Then move the blanket gradually toward and then into the base of the cat carrier: for some cats this may be achieved in a matter of minutes, for others it may take several training sessions. Aim to get to the stage where your cat will relax on the blanket in the base of the carrier (top still removed).

Now you can attach the roof. However, do this only when your cat is not in the carrier. If your cat is not keen to enter into the carrier now that the top is on, you may then need to go back a few steps, taking the blanket out of the carrier and rewarding him for being on the blanket near the carrier. Remember to shape the behavior—move the blanket farther into the carrier and reward closer approximations of the final behavior you desire. For example, your cat may initially place only his head in the carrier. Reward this behavior and gradually build up to head and one paw in the carrier, then head and two paws, then head, front paws and half of his body in the carrier, and so forth.

It can be awkward to present the food reward while your cat is facing into the carrier. You could try gently tossing it into the

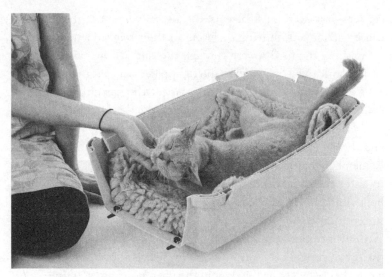

Sarah is rewarding Herbie's relaxed behavior with a chin scratch.

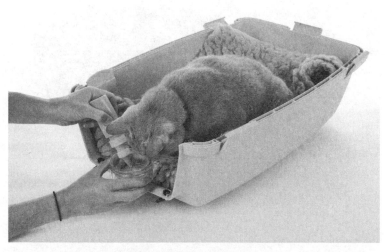

Sarah is rewarding Herbie's prolonged relaxation with a squeeze of meaty
paste into a glass dish from which he can lick.

carrier—however, for some cats, the fast movement of the food can cause excitement, moving them out of their relaxed state. Instead, try pushing the food reward through the slats in the carrier, which will allow it to be delivered to the right place without increasing the cat's arousal. Rewarding from this part of the carrier will also encourage your cat to go all the way in and stay there.

Our final goal here is to have the cat's whole body in the carrier and the cat showing relaxed behavior on the blanket in the carrier. Remember that at no stage have we had to touch our cat to get it to enter the carrier (unless we are using stroking as our reward)— he is unlikely to stay relaxed if he's being pushed in, however surreptitiously!

Only introduce the door of the carrier once your cat has stayed in the open carrier of his own free will over several training sessions in a row. Begin by fitting the door in the open position, at a time when he is not in the carrier. Once he is comfortable with being in the carrier with the door fitted but open for approximately five minutes,

Sarah is rewarding Herbie's relaxed behavior now that he is
shut in the carrier using some meaty paste on an elongated spoon.

begin to shut it, initially for only one second at a time, remembering to reward relaxed behavior and immediately afterward opening the door. Then, very gradually increase the lengths of time the door is shut.

It is important that your cat can cope with being in the carrier in a relaxed manner for the length of time of his longest journey. For example, if your vet trip is usually half an hour, see whether your cat can get to the stage where he remains in the carrier for this length of time of his own accord at home (with door open), and only then build up to similar lengths of time with the door shut. It is crucially important that your cat stays relaxed when in the carrier, as he will know that the shutting of the door means he cannot leave if he wants to. Feeding food rewards between the slats or through the door is a useful way of rewarding relaxed behavior in the carrier when the door is closed. You will now have a cat who is happy to spend time in his carrier in the home. Some cats can learn to enjoy being in the carrier so much that a new challenge emerges—how to get them to exit the carrier!

TRAVELING IN THE CAT CARRIER ADDS MANY NEW ELEMENTS beyond simply being in a confined space. Travel involves various kinds of motion—the carrier will be lifted off the ground and moved, and there will be some vibration during vehicle transit. Many cats find such movement hard to deal with, so it is really important to work in small steps. Start by ensuring your cat is relaxed in the carrier with the door closed, and gently move the carrier along the floor a few centimeters at a time without lifting it.

If your cat shows signs that he wants to leave the carrier at any time, perhaps by meowing or pawing the door, then immediately open the door and allow him to exit. We do not want him to feel any insecurity or loss of control at any time, as this will only hinder further learning. If this does occur, do not worry, it is just likely that your individual training goals were slightly too ambitious: you will need to break them down into smaller, more achievable tasks and work on them at a slower pace. It is also a good idea to reduce the time of each training session so that each session ends when your cat is still enjoying the training.

Once your cat has become accustomed to the carrier being moved gently around the floor, introduce the feeling of being lifted. Start by placing your hand on the handle and lifting the handle up without lifting the carrier off the ground—follow immediately with a reward for your cat. Repeat this several times—this will get your cat used to hearing noise above him and learning that such noise predicts a reward is on its way. If your cat is comfortable with this, proceed to applying a small lift each time you grasp the carrier handle—initially this lift should only be a couple of centimeters off the ground. Do not forget to give your cat a reward after every lift and to constantly check to see whether he is happy in the carrier.

Lifting the carrier solely by the handle to a height where you can comfortably walk with it can cause some carriers to wobble, which can be unsettling for the cat inside. For that reason, wherever possible, carry the carrier with both hands to stabilize it. Over several training sessions, you can build up the lifting until you are able to pick up the carrier and stand with it at waist height, just as you would if you were intending to walk off with it. Make sure both the lifting and setting down of the carrier is gentle and steady, as going into a box and being moved about is a strange sensation for cats, who generally like their feet firmly on the ground.

The next stage is to teach your cat that being moved around while in the carrier is nothing to fear. Start by simply walking a few steps, then stop and reward. Build this up gradually to being able to do short trips around the home with your cat in the carrier, each of which ends in a positive event. For example, you may ask your cat to enter into the cat carrier in the living room, close the door and carry him through to the kitchen where he can exit the cat carrier and immediately eat a tasty meal. Or you could carry him to another room where some of his favorite toys are set up, ready for a game with you.

The final stage is to introduce your cat to the outdoors in his carrier. Even if your cat does not usually go outside, he will have to experience some short outdoor trips in his lifetime; for example, even a trip to the vet's can involve four short trips outside—home to car, car to vet's, vet's to car and car to home. For an indoor-only cat with little experience of the outdoors, such trips are going to be really quite daunting, so the more practice you do—rewarding the trips with the best reinforcers for your cat—the easier your cat will find them.

If your cat does have access to the outdoors, you could use the carrier to take him to the garden and then let him out for a game or simply to explore if you already know that is something he finds rewarding. When you are ready to let your cat out of the carrier, open the door of the carrier slowly and quietly. Remember to provide a supply of rewards to your cat while he's still in the carrier—for example, small but tasty food morsels pushed through the slats or the door. Try to incorporate this training as part of your regular routine with your cat, practicing at least once a week—this way your cat is less likely to associate the carrier with unpleasant journeys, as he will have experienced more journeys that end in a positive outcome than those that end in something more daunting. Overall, this training teaches cats two things: first, being in the cat carrier is rewarding, and second, being moved in the cat carrier ends up with a positive destination or outcome.

The final step in cat carrier training is teaching your cat he has nothing to fear from traveling around in the carrier. Most common forms of transport with a cat involve motor transport by car or bus, or rail transport by train, metro, tram or tube. This is perhaps the hardest stage, and one that many cats find difficult. Therefore, it is important that you introduce travel not only gradually but also at as young an age as possible. The first stage is to teach your cat that being in your chosen mode of transport leads to lots of rewards. Before you start this stage, your cat must already be comfortable and relaxed in the cat carrier and while being walked around in it, indoors and outdoors.

If you need to take him by car, you already have one advantage because you can stay in control of when the car stops and starts.

Therefore, you can break the journey into smaller, more achievable subgoals for your cat to learn to cope with during the training stage. For example, when you first take your cat out to the car, its engine should be switched off completely, so there is no noise from the engine, radio, heating or air conditioning. Simply place the cat carrier on the seat in the car—the same seat where he will be once you actually transport him—and reward your cat. Remember to monitor your cat's behavior and body language at all times: as always, you should be rewarding relaxed and calm behavior. If your cat looks at all uneasy, remove him from the car and do more cat carrier practice in the home before trying again. One clear sign that your training steps are too large and your cat is not coping is when your cat will no longer touch the food treat.

The first few times you place your cat in the car, leave the car door open, and remove your cat (in his carrier) from the car after only a minute or so. You need to give your cat enough time to observe the interior of the car, but you do not want to leave him in there too long, otherwise anxiety may kick in. Provide a steady stream of rewards for relaxed behavior. Gradually build up your training goals, from initially just being in the car with you with the doors shut and the carrier belted into place on the seat with the engine off, to the same scenario but with the engine turned on. While the car is still stationary, let your cat experience some of the sounds and movements the car will make in transit—for example, the windscreen wipers moving, the indicators ticking and the gear stick moving. Do not perform these in succession, one after the other, but introduce one at a time over several sessions, always monitoring your cat and rewarding calm behavior.

The next step will be to go on very small trips in the car. If using public transport, short trips are usually the only way to start, because it is difficult to arrange to experience the train, bus or other form of public transport while it is stationary. However, it may be possible to simply visit the train station or a tram or bus stop without actually traveling by it. This will help get the cat accustomed to all the sounds and sights that come with this form of transport before having to experience the motion that accompanies them. If you are traveling by car and you usually drive, it is a good idea to have someone else drive during the early training sessions, to allow you to give your full attention to monitoring your cat and providing rewards. The first few trips should be brief—if in the car, a trip around the block or to the end of the street and back, or if on the bus or train, traveling one or two stops and back again at most. Although speed cannot be controlled on public transport, in the car it should be kept slow (less than 30 mph). Over time, build these trips up in length and in speed. Only once you are certain that your cat is fully comfortable in the car should you take over as the driver.

Now that you have a cat that is happy to travel, do not stop practicing. Keep up the regular trips that end with a positive event—for example, a ten-minute trip that ends back at the house with a tasty meal or game (see Key Skill No. 9). Make sure that your cat experiences more trips (in your favored mode of transport) that involve lots of rewards during the travel, with something good at the end, than trips that have more negative outcomes (e.g., an injection at the vet's). Remember to go at your own cat's pace, breaking down the training for the different components of the cat carrier and primary mode of travel into small and achievable goals. By doing this, you will help teach your cat to continue to perceive not just the carrier but also travel within it positively. This will greatly protect your cat's overall perception of travel from any negative incidents that happen to occur during your trips (for example, travel sickness—see nearby box). Furthermore, by gradually accustoming your cat first to his cat carrier and second to travel, you should find that your cat will cope much better with longer trips—for example, if you have to move house.

EVEN THE BEST-DESIGNED CAT CARRIER CAN COME TO HAVE unpleasant connotations for even the most laid-back cat, but it's not the cat carrier itself that's generally the problem—it's what has happened to the cat when he's been inside it. Cats instinctively like to go into quite small and apparently restrictive spaces, because they feel secure there, and their exceptional agility means that they know that if they have to, they can quickly make their escape. What they hate—perfectly understandably—is to feel trapped. Training a cat to go voluntarily into his carrier is therefore a matter of building on the positives—it's somewhere snug—and preventing the negatives— the carrier could be a trap—from coming out on top in the cat's mind. Another situation that cats sometimes love and other times recoil from is when they're touched. A touch can be enjoyable or a signal that something bad is about to happen: in the next chapter we'll see how to use training to maximize the former while minimizing the impact of the latter.

Troubles with travel

If your cat is struggling to feel fully comfortable traveling in the car, there are a number of other tips that may help the training.

- Some cats are frightened by the sight of the world whizzing past them through the windows: sun blinds or tinted windows can help reduce that stimulation.
- Some cats find the sound of the engine particularly unsettling. Playing gentle music such as classical music can mask the sound of the engine.
- Some cats experience travel sickness. If you have gone through all the training as advised, and your cat is still not traveling well, it is a good idea to speak to your vet, who will be able to diagnose whether your cat suffers from travel sickness and help treat this. Getting rid of the symptoms of travel sickness will greatly help your training—a cat cannot learn that the car is a positive environment if the movement of the car is making him feel sick. Furthermore, rewards such as food treats will not be tempting if he feels sick. Once he no longer feels sick, some of the training may need to be repeated to change his impressions of the car from negative to positive.

CHAPTER 8

Touch

Insult or indulgence?

O NE OF THE PLEASURES TO BE HAD FROM LIVING WITH A CAT is the sensation of stroking its soft fur. Scientific study has demonstrated that people obtain several benefits from stroking cats, some psychological, some physical. However, although many cats also appear to enjoy such interaction, for others, stroking can actually be a stressful experience. Fortunately, most pet cats (those who have had some early positive experiences of people) can be taught to enjoy being stroked. Although stroking generally occurs on mutual agreement between person and cat, there are other forms of physical interaction in which the cat may be less willing to engage voluntarily—these comprise health-care tasks such as grooming, particularly for long-haired cats, nail trimming and giving medications. However, if the cat is not comfortable being brushed or being restrained to receive medicine, he may start to avoid any situation involving touch: he may even start to avoid people completely. Training is the best way to teach your cat that the forms of touch needed for maintaining health, rather than being unwelcome, actually signal that a reward is on its way, and therefore become enjoyable in their own right.[1]

Touch is extremely important for a newborn kitten. Kittens are still blind and deaf when they're born, so touch, taste and smell are their only windows on the world. Physical contact with their mother is all-important to their survival, because not only is she their sole source of nourishment, she is their only source of warmth—kittens cannot regulate their own body temperature until they are several days old. As their eyes and ears open and they become more independent and self-reliant, so touch becomes less important. Nevertheless, while the litter is still together, the kittens will often all curl up in a heap, seemingly as much for reassurance as for physical comfort. And they will still physically greet their mother when she returns, often by rubbing their heads and flanks along her body (probably trying to persuade her to lie down and let them suckle from her, which as they get older she becomes increasingly reluctant to do).[2]

In the wild, some cats (especially unneutered toms) become loners as they get older, their contact with other cats restricted to fighting and mating. However, cats that have grown up together, females and neutered animals of both sexes, can maintain close friendships, reflected in periods of time that they spend resting side by side, often grooming one another. Sometimes such bonds will form between two cats that have not met until they were both adults. Although they more usually form between cats that have lived together for their whole lives, there are plenty of exceptions to this rule—these are cats, after all!

When a female farm cat rejoins her feline friends after a period of time away, perhaps following a patrol around her hunting grounds, she will usually approach with her tail raised upright (in some cats, the tip may also be curled to one side), and provided the others raise their tails too, they will all indulge in a bout of mutual rubbing, brushing their cheeks, flanks and on occasion, their tails on each other—sometimes by passing each other from opposite directions, sometimes by walking together side by side for a few steps. This is self-evidently a gesture of trust from one cat to another and apparently serves to reinforce the bond between them: it's often the younger or smaller cat that gets this process going, reflecting the probable origin of this ritual in the way that kittens greet their mothers. The physical contact is almost certainly important in its own right, but inevitably any smells that the returning cat has picked up on its travels will be transferred onto those he chooses to rub against. Whether this transfer of scent is of any significance to the cats themselves is unknown—it could serve to build up a common "colony odor," as is known to happen in some other carnivores, such as badgers. What we can be sure of is that our cats choose to greet us in exactly the same way, by raising their tails and rubbing around our legs. In doing this they seem to be acknowledging our greater size and the fact that we feed them, rather than the other way around.[3]

Cats that rest together often groom one another, and although this must help with keeping the harder-to-reach parts clean, it almost certainly has social significance, reinforcing the bond between the two cats. It may even help to restore relationships that have gone slightly awry, because it seems to be more intense immediately after two cats that normally get along well have had a mild argument. Also, unlike rubbing, adult cats do not restrict their grooming to cats that are larger or older than they are, implying that grooming is more of an expression of friendship between equals. (It's also true that mothers groom their kittens much more than vice versa, but this is primarily grooming for hygiene, not grooming for social purposes.)

When we stroke our cats, they often try to maneuver themselves so that we are stroking them around their heads, including the cheeks, around their ears and the backs of their necks. These are all areas that seem to be the number one targets for when one cat grooms another. Thus it's reasonable to suggest that cats find such stroking, at least on the head, neck and face, socially desirable, and hence also pleasurable. Pet cats also seem to like being stroked on the tops of their heads, especially in the area between their eyes and their ears where the hair is sparse and there is a concentration of scent-producing glands, so there may be something in the idea that they are encouraging us to take up their personal scent (sadly, a pointless exercise, because our insensitive noses are barely able to detect it). Most cats also like being stroked under their chins and around their mouths (another area of scent production), but very few like to be touched on their backs just in front of their tails (where a further set of glands can be found, although that may not be significant).[4]

There are other parts of their bodies where cats don't appreciate being touched. Their paws are not only very sensitive, they generally come into contact with another cat only during an altercation, so for one reason or the other, most likely both, cats tend to pull their paws away when we try to touch them. Most also dislike having their tails touched (despite the occasional involvement of the tail in mutual rubbing), possibly because it reminds them of incidents when their tail has become trapped or someone has tried to restrain them by hanging on to their tail as they tried to make their escape (not recommended). The soft underbelly is another place where many cats don't like to be touched, probably because they feel vulnerable to attack when they're on their backs. A few will tolerate a short period of this but may then hastily spring away, maybe even using a little too much force with their hind claws as they go. In this respect, cats shouldn't be confused with dogs, many of which adore having their tummies rubbed.

A cat that is relaxed wherever he is being touched will be easier to take care of than one that is willing to accept only a brief tickle behind the ears. Regular grooming is an important part of caring for any cat and if done right, not only is good for the cat's health but can also strengthen the bond between cat and owner. Cats naturally remove the dead hair from their coats when they groom, using the special hooks on their tongues, but by supplementing this with brushing, we can reduce the risk of too much hair being ingested and hairballs forming, prevent tangling of longer and thicker coats and remove any debris that may have been picked up on outdoor excursions, such as dust, dirt, cobwebs, small leaves and twigs. Even a shorthaired cat that lives exclusively indoors and grooms himself regularly can benefit from the occasional brushing, because it provides the ideal opportunity to check for external parasites, wounds and lumps and bumps.

Many cats naturally enjoy being brushed and need no formal training. Some cats do not mind being groomed around their head, neck and back areas but are not so keen on other parts of the body such as belly and legs, and some cats simply detest being groomed anywhere. How your cat reacts to the brush will largely depend on his previous grooming experiences and his tolerance for handling. However, with training, all cats can learn to enjoy grooming. For some, the feeling of being brushed may be intrinsically rewarding in its own right, or it could be that the cat learns to associate grooming with some other reward, such as food, and thus comes to enjoy being groomed because he knows it signals something tasty is on its way. Either way, once a cat has come to enjoy being groomed, it is an excellent way to spend some quality one-to-one time with your cat, allowing him to relax in your company.

GROOMING IS ESSENTIALLY JUST A FORM OF TOUCH, SO IT'S A GOOD idea to assess how tolerant your cat is of being touched even before getting the brush out. Because we will use our fingers to mimic grooming prior to the use of the brush, it is advisable to check, with your fingertips, whether your cat is comfortable being touched on his face, the top of his head and the back of his neck before commencing any training. Also, cats generally only groom themselves properly when they are relaxed, and so waiting until your cat is relaxed before introducing the brush will ensure he is at his most receptive to being groomed. You can also utilize the blanket on which you have previously trained your cat to settle and relax (see Key Skill No. 6).

Brushes come in various shapes, sizes and types; it is a good idea to use a brush with soft bristles when first introducing cats to grooming. The same is true for cats who have previously not enjoyed grooming and also applies to those with particularly sensitive skin or thin coats. For example, Herbie was an Asian cat and thus had a single coat. He was only ever brushed with a soft-bristle brush as this was enough to get through his coat without irritating his skin. Cosmos, however, is a domestic shorthair with a standard double coat, and the fine under-coat gets particularly thick in winter. He is brushed with a plastic-tipped, metal-pronged brush, and even with such a brush, Cosmos enjoys greater pressure being applied through the brush than Herbie did. Brushes that do not have protective plastic tips over the metal prongs should be avoided as they have the potential to scratch the skin. Likewise, combs should be avoided when first introducing a cat to grooming, as they do tend to drag the fur more than brushes.

Prior to introducing the brush to the cat, it can be a good idea to "load" it with your cat's scent so that when he does encounter it for the first time, it smells familiar and is thus less daunting. Do this us-ing Key Skill No. 7 to collect your cat's scent on a cloth and then rub this cloth onto the brush, which should be out of the cat's sight at this time.

For kittens, a soft toothbrush is ideal as their first grooming brush as it is not too overwhelming in size. Also, when moved in short, straight movements through the kitten's coat, it is likely to mimic his mother's tongue, thus helping to build positive associations with the grooming brush from the very first experience.

When we brush our cats, we should aim to mimic the actions that cats use when grooming each other. Some cats may even lick your hand or arm while you brush them: they are probably returning the favor of grooming, known as allo-grooming. Such behavior is a really positive sign, as it shows that your cat has a good bond with you and is enjoying the grooming for its own sake.

If your cat does not naturally find grooming pleasurable in its own right, the best thing you can do is to teach him that the action of the brush going through his coat predicts the arrival of rewards. At the same time, you should make sure that the cat always feels in control of the situation. In time, your cat should begin to find plea-sure in the sensation of being brushed, by association.

Start by placing the brush close to you on the ground and allow your cat to investigate it if he wishes. If he does investigate it, even if it is just a quick sniff or glance in its direction, reward such actions (see Key Skill No. 1)—remember, if using food rewards, tiny amounts

Before any training begins, Cosmos is given
the opportunity to investigate the brush.

are all that is needed, but make sure the value of the food is high
enough to be a real treat. At this stage we do not want the brush to
intrude into the cat's personal space, causing him to think about retreating if he does not feel comfortable—instead, it should be placed
at a comfortable distance from him, one that allows him to approach
if he wishes to. This way, he is making the decision whether to investigate the brush. If you follow this advice, your cat will feel more
in control and is likely to progress through the training goals quicker.

Once your cat is confidently investigating the grooming brush
while it is stationary on the floor, you can introduce a targeting
game with the brush. This consists of teaching your cat to follow the
brush, just as you would with a target stick (see Key Skill No. 3).
Remember to reward frequently and to remove the brush before rewarding, as this leads to the most efficient learning—placing the
brush behind your back is often a good way to remove the brush
from sight. A syringe or squeezy tube filled with meat paste can be

an excellent way of rewarding your cat for this task, as it can be kept behind the back in one hand when not in use and brought out quickly and easily for the cat to lick at the appropriate moment.

Once your cat will follow the brush for a few steps, you can then try holding the brush stationary at face height and see whether your cat spontaneously rubs his face on the brush. If he does, remove the brush only after he has finished rubbing on it (although you can mark the behavior when it occurs), and then reward him. Most friendly cats, new to the grooming brush, will offer this rubbing behavior relatively quickly (within one or two introductions to the brush). However, other cats may take longer to have the confidence to rub against the brush, usually because they have had previous negative experiences of grooming, perhaps because it has been painful or uncomfortable due to tangled fur, or they have been physically restrained to be brushed and thus felt trapped. This should not present a problem during training, so long as you work at your cat's preferred pace.

If after several sessions your cat is not rubbing on the brush spontaneously, you can hold the brush at your cat's face height, place your fingers between the brush and your cat and stroke your cat in his favorite place using these fingers, NOT the brush. Although he may perceive the stroking as a reward in itself, it is a good idea to reward him with something else (for example, a food treat) as well, as soon as the stroking has stopped, as the proximity of the unfamiliar brush may make the stroking less rewarding than normal. However, by associating the action with food rewards, your cat should soon be fully comfortable with being scratched with the brush in your hand. You can gradually build this up in terms of loosening your fingers around the brush so that some of the bristles touch your cat's fur, allowing him the opportunity to rub against the brush.

Once your cat is comfortably rubbing against the brush, you can gradually begin to move the brush across his cheeks, forehead and the back of his neck in the direction of the fur, using soft, slow movements. Initially try just one stroke before giving the reward, then build up to being able to carry out several brush strokes before stopping to reward (thus using the desensitizing technique in Key

Mutual grooming is one sign that your cat is comfortable with
the grooming process. Note at this stage only Sarah's fingers,
not the plastic coated bristles, are touching Cosmos's fur.

Skill No. 2). Meat paste delivered by syringe or squeezy tube forms
an ideal reward for this stage of training, as it can be delivered and
consumed quickly, allowing you to maintain momentum as you increase the number of brush strokes in each bout of grooming.

The order in which you present the brush and the reward must
always be as follows: reward only once you have removed the brush,
so that your cat learns that the brush predicts the arrival of the reward (Key Skill No. 5, touch-release-reward). If you introduce both
at the same time, he may focus his attention only on the food. Although it may then seem that he does not mind being brushed, it
may simply be that the food was enough of a distraction to allow

him to be brushed. Without food, he may revert to not liking grooming, showing that he hasn't truly learned that it can be a positive experience. So: brush first–remove brush–provide reward–remove reward–reintroduce brush, and so forth.

As always, remember to work at your cat's pace. If at any time he walks away or shows signs he is no longer comfortable or becomes too excited, give him a break and start another training session when he is relaxed again. Brief, frequent sessions are much more likely to succeed than one long training bout.

You may be tempted to groom your entire cat in one go, but most cats will find this too much—generally when grooming occurs between two cats, they focus on particular areas rather than grooming the entire body. Thus, be led by your cat and restrict each grooming session to just a couple of areas of his body. It is a good idea to always start with a few strokes of the brush at the face and head region before moving to other areas, as this mimics how cats greet one another and thus is likely to be the most acceptable form of grooming, from your cat's point of view. By grooming him in short bouts over many sessions, you will actually get more of your cat groomed than if you tried to tackle it all in one go. Moreover, your cat will always feel in control, helping to keep the whole experience positive. There are always exceptions, and some cats learn to adore being groomed and will happily let you groom for longer than just a few minutes. Of course, if you have such a cat, you can extend the grooming time.

As you progress through your training, move to different body areas on your cat in the order he likes them to be stroked. For example, after succeeding with the face area, it is probably a good idea to extend the grooming to a few strokes of the brush down the shoulders and along the back. Moving straight to the stomach area is likely to be a recipe for disaster, as this is the cat's most vulnerable area, and brushing there should be introduced only once your cat is fully comfortable being groomed everywhere else.

For some cats, nail trimming is an essential part of the grooming care package. Although many cats manage to keep their claws in tip-top condition without assistance, those who live an exclusively

Cosmos is being groomed gently.

After a couple of brush strokes, Cosmos is presented with a
syringe containing food as his reward for staying calm during grooming.

indoor lifestyle, as well as those who are elderly and less mobile, may
need help keeping their claws at a comfortable length. The first step
to claw trimming is to teach your cat to relax while having his claws
extended. The way to extend a cat's claws is to press gently with
forefinger and thumb on the toe: the claw should become visible
between the two furry pieces of skin that normally protect it. A cat
is generally most comfortable having this done when he is lying

down or, for those who enjoy it, lying in your lap (Key Skill No. 6, teaching relaxation, will certainly help). There are, therefore, numerous training goals to reach (see nearby box) before we even introduce the clippers. For this reason alone, not to mention that cats generally are very sensitive about having their paws touched, this is an activity that may take somewhat longer to train than some of the other health-care tasks.

Training goals to prepare for claw trimming

1. Touching front paw with forefinger
2. Touching front paw with forefinger and thumb (1–2 seconds)
3. Prolonged touching of front paw with forefinger and thumb (2–5 seconds)
4. Momentarily lifting front paw off ground/lap (1–2 seconds)
5. Lifting front paw off ground/lap (2–5 seconds)
6. Momentarily lifting front paw and extending leg (1–2 seconds)
7. Lifting front paw and extending leg (2–5 seconds)
8. Momentary gentle pressure on toe with paw lifted (1–2 seconds)
9. Momentary gentle pressure on toe with paw lifted and extended (1–2 seconds)
10. Prolonged gentle pressure on toe with paw lifted (2–5 seconds)
11. Prolonged gentle pressure on toe with paw lifted and leg extended (2–5 seconds)
12. Momentary pressure on toe to reveal claw (1–2 seconds)
13. Prolonged pressure on toe to reveal claw (2–5 seconds)

Nail clipping is a task that lends itself very well to the use of a meaty paste fed through a syringe as a reward, remembering this sequence: touch–withdraw touch–offer reward–allow cat to consume reward–remove reward dispenser (syringe or squeezy tube)–reintroduce touch. Toys and play are not suitable rewards in this training task, as we want to keep the cat nice and relaxed and certainly not encourage any paw-swiping. How many repetitions of each goal, how quickly you progress through the goals and how long you practice each specific goal are very dependent on your cat (1–2

Herbie is learning that gentle pressing on his toe to
reveal his claw predicts a food treat is on its way.

and 2–5 seconds are purely guidelines). Be guided by his behavior
and engagement and, of course, his level of relaxation. During this
task, we want a calm and relaxed cat throughout.

Once your cat is totally relaxed while having all his claws on a
single foot extended, you can introduce the nail clippers. If your cat
has had a previous negative experience with them, it may be worth
using clippers that look different (and preferably new, because these
will not hold any odors reminiscent of their previous use). As done
with the grooming brush, present the clippers near to you, but not
in your cat's own personal space, and allow him to come forward
and investigate them. You can reward him with a treat for doing
this. Then build up your training to gentle movement of the clippers
so that they lightly touch your cat's claw (keeping the clippers in
the closed position at this stage).

Before you actually use the clippers on your cat's claws, it is a good
idea to get your cat used to the noise they make, as this can sometimes startle or scare a cat. Systematic desensitization and counterconditioning (Key Skill No. 2) are crucial in nail clipper training;

Cosmos is learning that
clippers have no negative
consequences.

Sarah removes the clippers from Cosmos's foot
and rewards him with food using tongs.

you should accustom your cat first to the sight and then to the sound of the clippers, before any actual clipping starts. Clipping dried spaghetti or noodles makes a sound similar to that of clipping claws, so let your cat hear this sound, initially from a distance so as not to make him anxious. After each exposure, reward your cat so that he learns that this sound predicts a tasty treat is on its way. Through this learning—an example of classical conditioning—your cat will perceive the sound of the clippers as something positive. You will know that your cat has learned this when you see him orienting toward you after hearing the sound, in the manner he uses when he is expecting food—some cats meow, some purr, some circle and some simply stare at you. Initially, carry out this part of the training separately from training the claw touch.

Once your cat is comfortable with both components (touch and sound), they can be brought together, in small steps. Practice performing "pretend" clips, in which you go as far as applying pressure to a nail with the clippers but don't actually clip. By performing three or four pretend clips in between each real clip (every clip, real or pretend, should be rewarded), you will maintain the connection between the clippers and the food reward. It's a good idea to increase the size or quality of the reward for the real clips—for example, a greater squeeze of meaty paste from the syringe or the usual squeeze of meaty paste but adding a dried treat as well.

Do not set yourself the goal of clipping all your cat's claws in one go: be guided by your cat's behavior. In a single session, you may get only to the stage that your cat will happily accept one claw being clipped: don't regard this as failure. Several short sessions over several days—and a happy cat—are much better than trying to clip all the claws in one session, resulting in a very unhappy cat. If you are at all unsure how to clip your cat's nails or how much to remove, seek expert advice.

YOU CAN NEVER BE SURE WHEN YOU WILL HAVE TO MEDICATE YOUR feline friend. There will be some regular preventative medications such as flea and worming treatments, but there will also be unplanned treatments needed for injury or disease. Giving your cat medication does not need to be stressful for you or your cat, if you prepare your cat in advance by undertaking some simple training and repeating it every so often so that your cat doesn't forget.

You may ask why you should spend time training, because you can just hide a tablet in food or physically hold your cat until he swallows the medication. However, these tricks often work only temporarily, until the cat catches on and begins to eat the food and then spits out the tablet (a perfect demonstration of how fast cats can learn when they feel the need to). If you have to restrain him, he will increasingly struggle during restraint.

Furthermore, so often the only time a cat receives the handling required to give medicine is when he is already ill. At such times, cats are already likely to be in some level of discomfort or pain, so will come to associate your handling (and the treatment) only with this pain or discomfort. Therefore, it is a good idea to start training when your cat is in good health and can learn to associate the handling and treatment with something positive. Teaching your cat to accept medication through fun and games is an excellent life-time skill.

Some medications that need to be swallowed can conveniently be given by syringe, allowing you to feed liquids or powdered pills easily. Some capsules can be opened and the powdered content mixed with food or liquid, and then can also be fed via a syringe. However, we recommend you seek advice from your veterinary surgeon before manipulating any of your cat's medication.

Because many of our training tasks incorporate the use of paste-like food rewards provided through a syringe or on a spoon, your cat should already consider this a positive event. When choosing a food to add the medication to, make sure you select a strongly flavored but tasty mixture to help mask the taste of the medication. Start by offering the syringe or spoon several times with the chosen tasty food on its own, without the medication, and only then add the medication. Continue to offer the unmedicated food several times between each medicated dose, so that you don't break the association between being fed by a syringe or spoon and getting something really tasty to eat.

If using a syringe, remember not to press the syringe plunger down too quickly and also not while the cat is actually licking from it—simply press very slowly for a second or two so that some food comes out of the tip and then present that to your cat, allowing him to lick the food from the syringe tip at his own pace. Once your cat is confident with this procedure, you can start to very slowly apply pressure on the plunger so a little more food comes out while your cat is still licking (but be careful not to scare your cat). Keep each session short so that you finish leaving your cat wanting more, rather than having to stop because your cat has lost interest and walked away.

You may find that your cat licks the syringe or spoon initially but then gradually refuses to take any more. This may be because he can taste the medication: even if it is not actually disgusting, it may have changed the taste of his regular food enough to turn him off. One way around this is to provide a reward after each lick: that reward could be an even tastier piece of food such as a small piece of fish or chicken or something else that your cat really values, such as a game with a toy or a gentle fuss. Remember, some cats enjoy attention, praise and petting as a reward: if so, you can confine how much of this you give them during a training session to just the few seconds after your cat has accepted some of the substance with the medication in it. With the addition of a reward for licking the syringe or spoon, you should be able to increase the amount of the syringe or spoon contents the cat will consume before receiving his chosen reward. Several repetitions of offering the syringe or spoon and then following up immediately with the chosen reward will teach the cat that licking from the syringe or spoon predicts something he really likes.

If your veterinary surgeon advises that the medication you have is not suitable for crushing and placing in a syringe with a liquidized food, your cat will need to accept it whole. This is generally something cats do not like at all, but with a few simple training steps, done well in advance and preferably not when the cat is ill, you can make sure that this will be a much less unpleasant experience for your cat. Choose a time when your cat is relaxed and resting to begin training—you may wish to utilize your cat's special relaxation blanket to help keep him calm and relaxed throughout.

Ensuring your cat's relaxation blanket keeps its association with relaxation

The more you utilize your cat's relaxation blanket (Key Skill No. 6) in other training tasks—for example, training your cat to accept medication—the more likely the blanket will begin to lose its association with relaxation, because in the initial training of such tasks, the cat may well feel a little uncomfortable. Thus, it is really important to keep shaping your cat to relax on the blanket at times where you do not require its use for another training task—this will maintain the blanket's power to help your cat to relax. Recall that shaping involves the rewarding of successive approximations of the goal you have: in this case, our goal is relaxation. Thus, we reward any postural changes that indicate the cat is settling down to rest, as well as any behavior that shows his mood is relaxed.

If you first teach your cat to relax on the blanket but then place a new, potentially daunting task in front of him every time he is on the blanket, he will eventually no longer associate the blanket with relaxation. We can therefore think of our regular sessions of rewarding relaxed behavior on the blanket as recharging the blanket's relaxation potential.

The first goal of the training is to be able to place one hand around your cat's head, with your thumb and index finger placed against either side of your cat's top lips. This position will be required to open your cat's mouth to place the pill inside. How quickly you reach this goal and how many intermediate goals you devise to get there is very much dependent on how much your cat likes being touched. For example, if your cat is not so keen on this, start with the aim of simply touching the top of his head, and reward him while he is calm. If you already know that your cat enjoys being touched on the head area, an ideal first goal can be having your cat accept your hand placed over the top of his head.

Once your cat is comfortable with a hand placed over the top of his head, the next step is to teach him to be comfortable with the finger and thumb of your other hand gently touching his lower jaw. For most cats, this will be a new sensation, so it is important to work slowly and gently but purposefully. Your cat is likely to sense lack of confidence in your handling, and he may then draw away from you. Provide plentiful rewards after each touch. If your cat flinches at all, give your cat time to settle and relax before trying again.

The next goal is to teach your cat to lower his bottom jaw when touched, so as soon as he lowers it, even to the smallest extent, make sure you are ready with your rewards. Most cats do not enjoy opening their mouths for examination—the way to turn this around is regular practice in short sessions (simply touching the jaw once or twice in each session to begin with) and providing rewards your cat really likes. When you feel your cat's jaw loosen slightly as you touch it, use both your hands to open it slightly for a split second before releasing and rewarding. Build this up over several sessions until you can hold your cat's mouth wide open for a second or two. Cats generally do

Teaching Herbie to cope with having his head
held still as part of learning to take a tablet.

not like to open their mouths for very long, so aim to make this process as quick and positive as possible.

If you find that your cat tries to move away, allow him to do so: you have probably asked too much from him too quickly. It would be advisable to ask for something slightly easier when you do your next training session.

Next, you should practice holding his jaw shut for a second— your cat will most likely naturally shut his mouth after you release the opening pressure. You will need to do this after you place the tablet in his mouth, to help your cat swallow the tablet. Remember, teach in small steps.

The final step is to get your cat to accept the tablet. Once your cat is comfortable opening his mouth for a couple of seconds for you, progress to holding a tablet between your thumb and index finger of the hand that you will use to lower the jaw (you may need to move to opening the lower jaw with your middle finger). Find whatever position is most comfortable for you and your cat. Do not yet try to place the tablet in the cat's mouth, just practice the opening of the mouth as before, the difference being that you are now holding a

Herbie learns to stay calm while having his mouth opened.

tablet. Although you may not think of this as a big step, your cat may feel differently.

Once your cat is comfortable with a tablet being held in front of him, you can try placing the tablet into the back of his mouth, then gently closing the mouth as practiced above. The secret to getting your cat to swallow the tablet is to place it as far back in the mouth as possible. Some tablets do not taste nice to cats, and they use their tongues to bring them to the front of their mouths to spit them out. By placing the tablet as far back in the mouth as possible, the cat is more likely to swallow the tablet rather than spitting it out: pushing the tablet back into the mouth using your index finger can help. As soon as you feel your cat swallow, reinforce highly and generously with a very high-value reward. Use something that is tasty and also wet in consistency, such as meat in gravy, as this type of food encourages the cat to swallow freely, preventing the tablet from being stuck in his throat. Additionally, you can rub or scratch the throat area (but only if your cat enjoys this), as this also encourages swallowing.

If your cat manages to spit the tablet out, it is still important to provide a reward after placing the tablet in the mouth, as your cat

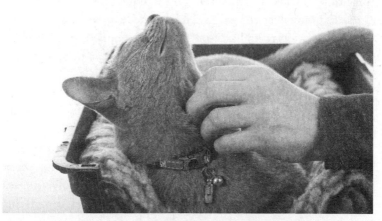

Herbie enjoys a scratch under his chin as a reward for
accepting his tablet—this action also encourages him to swallow it.

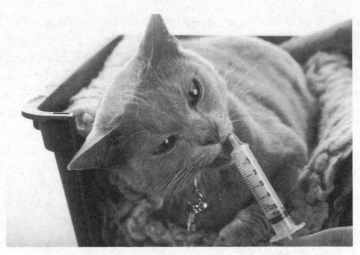

Some syringed meat paste acts as a high-value reward for Herbie
as well as encouraging him to swallow after accepting his tablet.

did open and close his mouth for you as asked. Try to stay calm and
patient, as your cat will probably react adversely to any sudden or
rough movement, and will be less keen to repeat the procedure in
the future. Do not immediately repeat the task but give your cat
some time to relax, and then try again at a quiet time.

Apart from tablets, common classes of medication for cats include
eye drops, ear drops and spot-on treatments applied to the back of
the neck. The principles behind training your cat to accept such
medications voluntarily are just the same as for tablets.

The first task is to break the end goal down into small successive
steps, and that starts with imitations of medication (the systematic
desensitization part of Key Skill No. 2), which the cat is rewarded
for accepting without fear (the counterconditioning part of Key
Skill No. 2). Although cats should never be given medication when
they do not need it (and therefore training cannot involve the actual administration of the medication), every other step toward this
end goal can be practiced and taught to be predictive of rewards.
Such steps include being held in a particular way (for eye drops, the

eyelids held open; for ear drops, the outer ear folded back to reveal the ear canal; for spot-on treatments, the parting of the hair at the back of the neck) and experiencing new sensations (for example, administering a spot-on treatment involves the liquid medication being placed directly onto the skin—we can mimic this with a small amount of water).

Remember to allow your cat to investigate any object the medication comes in before attempting to place it on or close to your cat's body (Key Skill No. 1). Allowing your cat to see and explore the medication container before you use it gives your cat some feeling of control over the situation. Then, reward your cat after holding the container for the medication against him in the way you will when medication is actually being given, with the unopened medication in front of you. This should lead to a situation where the cat is much more likely to cope when the medication does need to be given.

Positive practice training experiences such as these should increase your cat's trust and tolerance of handling and administration of actual medication. Ultimately, with any health-care training, and

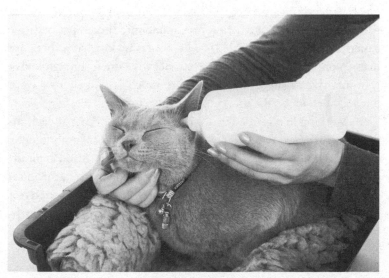

Herbie calmly accepts the sensation of the bottle's nozzle.

indeed any training of our cats, we are not teaching them to be obedient and meet our every demand. Instead, we are training them to cooperate voluntarily and to trust us when we present new experiences. We should always be working within their limits and adjusting our expectations to fit these, rather than asking or expecting too much and ending up with feelings of failure and despair on our part and feelings of anxiety, fear or frustration from our cats.

The key to success for training all health-care tasks is therefore to break down each task into small achievable steps, to prevent the cat from ever feeling any worry. Different aspects of the training goal may be separated into individual subgoals and trained in their own right before bringing these sensations together. For example, you may work on the sight, sound and feel of some health-care equipment against the body independently, only bringing each aspect together when the cat is fully comfortable with each individually. Aim for several short sessions each month. Leaving training until a health-care procedure cannot be avoided will only add pressure on yourself to succeed, and you will be tempted to run through the training goals too quickly, with too few repetitions. Sadly, it is much easier to create a negative association than a positive one, so patience and plenty of practice are paramount. Sometimes negative experiences are unavoidable, but by always making sure there are many more positive experiences than negative ones, through lots of training, the negative experiences are much less likely to have a long-term effect on the way your cat reacts to you.

By working in a gentle but systematic order, working at each individual cat's own pace, you will find that this training philosophy can successfully be used to get your cat accustomed to and comfortable with all types of health-care procedures ranging from the mundane of grooming to more specific cases such as insulin injections for diabetic cats and accepting medication from an inhaler for asthmatic cats. Such training will also help set up a solid foundation for the training we focus on in the next chapter—extending this acceptance of health care at home to the veterinarian's office.

CHAPTER 9

Flight, Fight or Freeze?

Cats don't take kindly to stress (especially visits to the vet's)

CATS RELY ON THEIR INSTINCTS TO KEEP THEM OUT OF TROUBLE. Unfortunately, what they perceive as "trouble" can sometimes be something that's being done for their own good. The example that is most likely to come to all owners' minds is a visit to the veterinarian. For a cat that hasn't received any training, a visit to the vet's can be one of the scariest events in his life. Not only does he have to contend with going in the carrier and traveling by car; he then often has to wait in a noisy room full of people, cats, dogs and possibly other animals that are not only unfamiliar to him but, in his current state of mind, also potentially threatening. In novel situations, a cat's threshold for coping with unpredictability is greatly lowered. Thus, any attention from any unknown human or animal is most likely unwelcome. Even if attention is not focused directly toward the cat, the constant movement of unfamiliar people and animals going past the carrier, coupled with being unable to retreat to a safe distance, is likely to create increasing anxiety.

Once inside the consultation room, the cat is suddenly presented with the opportunity to leave his carrier. However, when faced with the open expanse of the consultation table, which certainly smells unfamiliar and probably also repulsive due to the disinfectant (recall that a cat's sense of smell is many times more sensitive than ours), the cat usually perceives that the safest thing to do is remain within the carrier—indeed he may freeze. Also at one end of the table is the veterinarian, whom the cat may be frightened of, for not just one but two reasons. First, the cat may not know that particular vet, and unknown people, particularly when encountered in a new place, can be daunting to a cat. Alternatively, the cat may know the vet but recalls that previous experiences with her have involved pain or discomfort, thereby creating a negative association in the cat's mind of vets in general. Either way, the poor cat is by now already very cautious and on edge. If he does not voluntarily exit the carrier, the veterinarian may reach in and pull the cat out: however gently she does this, the physical loss of control and inability to hide is only more likely to lead to the cat feeling helpless—his stress response will now be at full throttle. Thus, by the time the cat is faced with a physical examination, he may be frozen in fear or desperately struggling to get back to the protection of his carrier or the underside of a cupboard. Even cats that dislike their carrier may find it suddenly their best option in the face of unfamiliar or unpleasant veterinary procedures.

This problem might not arise if we had any way of letting our cats know that a visit to the veterinarian is ultimately for their benefit— either immediately (for example, to fix up an injury or medical problem) or in the future (for example, to receive a vaccination that will protect him from disease). Cats live very much in the present, and the concept of "endure now, benefit later" is beyond their grasp (indeed, it's hard enough for many of us to be entirely rational when faced with such dilemmas). It's therefore fruitless and ultimately counterproductive to wish that our cats were instinctively a little more trusting of us. It's far better for them—and for us in the long run—if we accept that we may need to teach our cats that certain situations, specifically those that we know they will naturally react negatively toward despite their being in their best interest, are in fact OK, even mildly pleasurable. The route to doing this is, of course, through simple training, first reducing the stimuli emanating from the vet, and the handling, procedures and equipment she utilizes, to intensities the cat can cope with, and then associating the same stimuli with pleasurable outcomes—systematic desensitization and counterconditioning (Key Skill No. 2).

Unlike dogs, cats rely on a basic set of instinctive actions when they feel threatened. Dogs, who are intensely social animals, will usually turn to their owners for reassurance—sometimes hiding behind another dog will serve as well, but the principle remains the same. Cats retain a strong feeling of self-reliance from their wild ancestors, and because in the wild the latter had plenty of enemies—humankind included—most of today's cats still tend to run away from anything they perceive to be a threat. If many cats seem to overreact to new situations, objects and animals, this is understandable in terms of their evolution. Wildcats that respond to a hundred false alarms will have wasted some hunting time, but on each occasion once the interruption is over they can pick up their lives where they left off; any wildcat that failed to respond to a genuine danger just once might never be given that chance again. Pet cats will eventually habituate to some things, for example, noises such as a telephone ringing that incur no relevant outcome as far as they are concerned, but if they sense danger, they may go the other way—overreacting even more (recall that this process is known as sensitization). So any cat who recalls previous encounters with the veterinarian in a negative way (either due to pain, fear or both) will be much less likely to habituate and may indeed sensitize, so that subsequent visits will be stressful for the cat, even if no pain is experienced at that time.

JUST AS A MOUSE RUNS FROM A CAT, SO A CAT WILL PREFER TO flee from danger. Sometimes "danger" arises from pure instinct—for example, the cat's first encounter with a large and cat-averse dog. On other occasions the cat's perception of "danger" arises from something he has learned, say the arrival of a stethoscope if his previous experiences of being examined by a veterinarian have not been pleasant. In both instances the cat's reaction is simple and standardized—put some distance between yourself and where the problem appears to be. What the cat does next will depend partly upon his personality and partly on how bad he perceives the threat to be. At one extreme, the cat may struggle loose from the vet's grasp and dart to the nearest hiding place (often the cat carrier) and refuse to come out. At the other extreme, the cat may move across the veterinary table and then stop and turn his head to examine the threat from what he perceives to be a safe distance, while remaining poised to resume his rapid exit.[1]

Sometimes a cat will feel that flight is not an option, perhaps because he is being physically restrained, as is common practice during a physical examination, or because he can see nowhere safe to run. He may then resort to confronting the source of danger, growling, hissing, spitting and even scratching and biting. Cats that have repeatedly felt under threat in the past may even use a brief burst of aggression as their first reaction, attempting to flee only once they have delivered a painful warning. Such cats are rarely favorites at the veterinary clinic! However, it is important to remember that such aggression is born out of fear (and possibly frustration at being unable to escape), not anger—and certainly not wickedness.

The third option available to any animal in danger is to freeze. Presumably by remaining motionless the cat hopes to fool its "enemy" into failing to notice it, or at least to appear completely unthreatening. Some mammals specialize in this kind of response: rabbits, for example, sometimes "play dead" in front of a predator if they reckon that escape would be futile, hoping to fool their attacker until its attention wanders for a moment, providing a brief opportunity for escape. In pet cats, freezing is less common, and when it occurs is usually thought to be a reaction to repeated, that is, chronic, stress. Such apparent inability to move is often termed *learned helplessness* (see nearby box). For example, some cats hate to be caged, but finding that nothing they do will allow them to escape, switch to learned helplessness if they have had to be confined several times or for a prolonged period of time, such as several days or weeks after a hospitalization. A cat in this state may appear relaxed to the casual observer, but in fact he will be anything but.

The key is to be sensitive to subtleties of the cat's body language. The position of his front paws is one of the key features in distinguishing between real and apparent relaxation in a cat that is lying down. Cats that are genuinely relaxed will tuck their paws underneath their chests, rotating their paws so the pads are off the ground, or will splay them out while flopping onto their sides. Cats that are tense will "rest" with their front paws visible in front of them, and all four sets of pads in contact with the ground, ready to spring up at a moment's notice. Chronically stressed cats may even close their eyes, giving the impression that they are "asleep," but their twitching ears, the tenseness of their muscles, and their tightly pressed eyelids all betray their true state of vigilance. Furthermore, when a tense cat is approached, you may see his body drop even closer to the ground, restricting your visibility of his paws: his eyes and pupils may widen or the eyelids may tightly squeeze shut, his head may be pulled in closer to his body and lowered, his ears may appear to flatten, and his tail will be clamped tightly against his body. Such a cat is doing everything in his power to avoid physical contact with the approaching person.

Whatever the reaction, within a few seconds of his perceiving danger, the cat's heart will be thumping and the amount of adrenaline (epinephrine) in his bloodstream will be rising. This hormone is known to enhance feelings of fear and to stimulate the learning of negative associations. Thus this is not a good moment to start or even continue with training, otherwise the cat may make the wrong associations.[2]

From a training perspective, a stressed cat, regardless of whether that stress is short lived or chronic, is unlikely to learn much that is useful. Indeed, he is most likely to learn ways that he can reduce his fear, anxiety and frustration for himself, either by running away, freezing or becoming aggressive, none of which is good for the cat in the long term, and especially not for his relationship with his owner. Four points should be noted:

1. Never try to train a recently stressed cat: his attention will still be occupied by the perceived threat (whether real or imaginary), and not on the task you want him to learn.

2. In order to help alleviate a cat's stress, give him space, quiet and a place to hide.

3. A recently stressed cat will let you know when his stress levels have reduced enough to commence training by leaving his hiding place.

4. Training using rewards is the best way to reduce, and eventually eliminate, your cat's perceptions of threats that are more apparent than real—say, the arrival of the veterinarian into the consultation room, the smell of a hospital cage or being held still for a veterinary health check.

Stress is bad for cats—right?

We use the word "stress" to describe a conscious experience of mental or emotional strain: Do cats experience something similar? It's impossible for us to get completely inside the emotional world of cats, but we do know that they are capable of experiencing something akin to anxiety (a general feeling of fear), as well as the more straightforward (and obvious) fear of something that's right in front of their noses. Furthermore, recent research has shown that cats can experience frustration, another form of stress, when something they desire is unobtainable (i.e., escape) or their expectations are not met. It is also becoming increasingly apparent that prolonged stress can not only make a cat behave oddly but may also even contribute to such medical conditions as cystitis and dermatitis.[3]

The question that therefore begs to be asked is, if stress is so bad for us—and for cats and for all other mammals—why hasn't it been selected out by evolution? The answer is that stress is actually essential to survival and only turns into a problem when it becomes continuous. Unfortunately, crowded modern living conditions are making stress increasingly more common in cats (as well as in other domestic animals and people).

The stress reaction in itself is a vital mechanism for getting out of trouble quickly. Pausing to think, "How dangerous is this, or is it even dangerous at all?" will lead to a vital second or two delay, which might prove fatal. Hence, most animals (and humans) rely on an inbuilt set of

(continues)

Stress is bad for cats—right? *(continued)*

responses that take them far enough away from trouble to give them enough breathing space to evaluate the situation.

Our bodies—and cats'—have two distinct mechanisms for reacting to danger, one more rapid than the other. Immediately as the threat is perceived, the hypothalamus, a structure at the base of the brain, is activated, leading to a burst of the hormone adrenaline (epinephrine) from the adrenal glands (attached to the kidneys). This prepares the body to deal with the immediate threat: the heart speeds up and breathing becomes faster and deeper, pumping energy and oxygen into the muscles in preparation for a life-preserving sprint or for fighting if escape is impossible.

If the danger continues or is repeated (for example, a cat gets trapped or lost), then a second system is activated, releasing a different hormone, cortisol, from the adrenal gland. This speeds up the release of glucose into the blood, reducing fatigue and also decreasing swelling after any injury. Nonessential activities such as digestion of food are suppressed, returning to normal once the threat has passed and the amount of cortisol in the blood has stabilized.[4]

All of this obviously should help any animal—including ourselves—to react appropriately to a sudden but transient threat. However, serious problems arise if the threat continues. If cortisol continues to enter the bloodstream, it will eventually suppress the immune system and render the cat more susceptible to infection, not to mention autoimmune disease.

Cats differ in how they show that they're chronically stressed. Some will become very active, possibly pacing and meowing incessantly; others become withdrawn and apparently unreactive. Which response is shown depends on the individual cat, how it perceives the context it finds itself in and whether it is experiencing anxiety, fear or frustration. Helping a cat that has become chronically stressed requires more than just training and so is outside the scope of this book.[5]

Luckily for most pet cats, visits to the vet don't have to be filled with fear and anxiety. As knowledge of cats increases, in recent years we have seen movements toward changing the way cats are managed in many veterinary practices, with programs being implemented to help cats feel more comfortable. One such example is the

Cat Friendly Practice Scheme. But however cat friendly a veterinary practice may be, training your cat to cope with a veterinary visit will allow your cat to receive the best veterinary care in as stress-free a way as possible. For example, the more your cat is able to cope with handling by the vet, the less restraint he will need—a major benefit, because heavy restraint is unpleasant for both your cat and the veterinary staff, and it can be distressing for you to see your cat so frightened. Furthermore, with heavy restraint removed, the veterinarian can be more confident that any negative responses shown by your cat during examination are related to pain or discomfort associated with injury or disease, rather than fear, anxiety or intolerance (frustration) of being handled and examined. Thus the vet can be more confident of his or her diagnosis.[6]

It is easy to see that visiting the veterinary clinic encompasses many key skills and many training exercises we have covered in previous chapters (i.e., meeting new people, going into and traveling in the cat carrier, being touched by you and by strangers, and healthcare procedures carried out at home). Thus, before you begin training in preparation for veterinary handling, it is important that your cat has already mastered these training exercises, because they will provide a solid foundation for learning that a visit to the veterinary clinic is nothing to fear.

IT CANNOT BE EMPHASIZED OFTEN ENOUGH THAT CATS DO NOT learn well when they are stressed. Removing or diminishing as many of the things at the vet's that cause your cat to feel stressed will make training your cat for a visit to the vet easier. Although many of these things may be in the veterinary clinic's control, it is important that you are aware of what they are so that you can choose the best clinic for you and your cat and make best use of the available services.

Minimizing exposure to other animals in the waiting room will help prevent any stress your cat may be feeling from getting worse. Avoiding dogs can be achieved by attending feline-only clinics or by asking to be allowed to wait with your cat in a dog-free section of

---

Cat-friendly features in a veterinary clinic

Waiting Room

- Cat-only waiting areas
- Cat-only clinics
- Raised areas for placing cat carriers
- Seclusion for individual cat carriers
- Reception desk containing raised place to position cat carrier
- Cat treats available

Consultation Room

- Dedicated feline-only consultation room
- Non-slip surface on veterinary examination table
- Examination conducted within the base of the cat carrier
- Towels available to make a nest for the cat to hide in during the examination
- Cat treats available
- Soft, gentle handling by all staff

Hospitalization

- Cat-only wards
- Cages above ground level
- Hiding place provided in veterinary cage

---

the waiting room. If your veterinary practice offers neither of these, find the quietest spot in the waiting room, or even ask the receptionist if it is possible to wait in the car and be notified when the veterinarian is ready to see you.

Cats gain security from being high up, so placing the cat carrier on a designated shelf or perch, or if none is available, on your lap, is preferable to leaving the carrier on the floor. Likewise, cats feel safer when they are partially concealed, so bring familiar bedding such as a blanket and drape it over the carrier in such a way that your cat can choose whether to be in view or to hide. By enhancing your cat's perceived security and reducing his exposure to other animals, he will have a better chance of remaining calm and being in a position to see that his safety is not in danger.

The best way to prevent your cat becoming fearful in the waiting room is to have already exposed him to the waiting room in ways that were positive. Building up a bank of positive experiences can help your cat to view future visits as unthreatening: the more positive experiences your cat has received, the greater the protection against a negative experience. One way we can build up positive experiences is to ask the reception staff whether it is OK to bring your cat into the waiting room for a short period during their quieter times—that will allow your cat to experience the waiting room when it is least daunting and teach your cat that such visits are not automatically followed by (i.e., do not predict) negative events.

It is important when giving your cat the opportunity to learn positive associations that he remains within his comfort zone throughout and be rewarded while in this emotional state. Because your cat will be in his carrier in the waiting room, physical interaction such as stroking can be difficult beyond a couple of fingers through the door to scratch under his chin or behind his ears. However, food rewards can be easily delivered into the cat carrier, as can the opportunity to play with a toy—a wand toy can be used with the wand outside of the carrier and the toy inside.

Your first visit to the veterinary waiting room may be for only a few minutes—long enough for your cat to observe the environment but short enough so that he remains calm and feels in control. Subsequent visits can increase in time, so long as your cat remains calm and relaxed. How many visits you need before your cat gets a positive impression of the waiting room very much depends on the individual cat—a confident, sociable kitten with no previous experience (or at least no negative experiences) of the veterinary clinic may only need one or two visits to create such an association. At the opposite end of the spectrum, a less confident cat with bad memories of the vet's may need more visits to change the negative association to one that is positive. Likewise, such a cat is likely to be able to cope with only short visits at first and will take longer to feel comfortable enough for visits to increase in length. Thus, starting such training from kittenhood is ideal. Veterinarians will be most grateful when they meet a confident, happy cat on subsequent visits, as opposed to the typical fearful cat threatening to use his claws and teeth.

Extending the idea of building up positive associations with the waiting room, you should also try to expose your cat to as many positive contacts with veterinary staff as you can. For example, you can ask the reception staff to feed your cat a treat through his carrier door. Doing so will help prevent the more negative encounters (e.g., a vaccination appointment) clouding your cat's positive view. Most people (understandably) take their cats to the vet's only when there is something wrong or the cat needs preventative health care such as booster vaccinations. Because these are most cats' only experiences of the vet's, and all are likely to have involved some element of discomfort and pain, it is easy to see how these cats begin to fear veterinary visits—all have contained a negative event.

More forward-thinking vet practices should be willing to help you to get your cat to relax while waiting for treatment. Many now offer the opportunity for your cat to meet with a veterinary nurse for a basic health check, where gentle handling, play and food treats are on offer to give your cat the opportunity to build up positive experiences. Some even offer familiarization visits where no health check is performed but the cat is simply given the chance to explore the consultation room in his own time, building an association of the surroundings and especially the veterinary staff with rewards, such as food treats, play and gentle stroking (if the cat desires it). Generally, practices target these visits specifically at kittens, thus allowing cats to make such positive associations from an early age. However, do not be afraid to ask your practice whether they can extend such visits to your adult cat if he needs some training to be relaxed on their premises. They ought to react positively, appreciating your proactive approach to helping your cat: ultimately, it will make their job of looking after your cat easier if he is amenable to being examined and treated.[7]

As well as preparing your cat for visits to the vet's premises, it is also a good idea to introduce him to approximations of many of the veterinary examination techniques at home. For example, we can teach cats to learn that having their eyes, ears, skin and teeth looked at all predict that rewards are on their way. We can also practice supporting our cats in a standing position, which is often used in the veterinary practice, and although it is not a good idea to fully palpate your cat's belly area—leave this to a qualified veterinarian—we can teach our cats to be comfortable with having this vulnerable area touched.

Having the opportunity to teach your cat that such handling predicts reward in the calm, quiet and stress-free environment of the home, and at a pace he is entirely comfortable with, will greatly boost his ability to cope with an examination when it occurs during the veterinary consultation. Such experiences will no longer be entirely novel nor potentially frightening, but will be familiar and associated with reward. The only difference will then be that at the vet's they are being performed in a more challenging environment, but ideally by this point your practice visits to the vet will have greatly reduced your cat's perception of threat. Although you cannot always guarantee that your cat will be seen by a veterinarian he has previously met and has positive associations with, your previous training centered on teaching your cat to find visitors to the home rewarding should also help eliminate any anxiety here.

Dummy examinations at home can take place on the conditioned relaxation blanket (Key Skill No. 6) in the base of the cat carrier: your cat will feel nice and relaxed here (if you have carried out the training detailed in Chapter 7), and thus he will be in the ideal emotional state. Furthermore, encouraging your vet to perform physical examinations in the base of the cat carrier will give your cat some additional perceived safety and security, rather than feeling exposed on the open expanse of the veterinary table.

To teach examination of the teeth and mouth, begin by scratching under the chin and cheek area—places where cats tend to prefer to be touched anyway—and then move to gently touch the lip,

Herbie is learning to have his teeth examined—starting
with a gentle pull back of the lips.

removing the fingers and presenting the reward in the manner explained in Key Skill No. 5 (in Chapter 3). Repeat this several times,
starting with just a touch of the lip and then progressing to a gentle
lift of one side of the top lip to expose an upper canine tooth. If you
need to hold your cat's head in your other hand, begin to introduce
this by touching the head and the lip at the same time. Remember,
gentle handling and taking small, successive steps always followed
by reward is the way to progress.

   If at any time your cat displays body language indicating discon-
tent (e.g., slight flattening of the body, ears or both, turning the
head away from you) or increasing arousal (e.g., dilating of pupils,
tail swish), stop immediately. Your cat is telling you he is not enjoying the experience, and thus what you have just done is a step too
far. Wait until another time when he is relaxed, and start right back

from the beginning. Over several sessions, each of short duration, advance your training to the point where you can separate both top and bottom lips to see the teeth at the front of the mouth, as well as being able to gently pull the part of the lips that create the curve of the mouth at the cheeks to see the back teeth—just to the point where you can look at your cat's teeth for a second or two. Using the same training techniques, you can develop positive associations with having the ears touched and pushed back to inspect inside, with having eyelids gently widened to examine the eyes, with having the fur brushed in the opposite direction from the growth with a finger to reveal the skin, and with placing your hand under the belly to hold the cat in a standing position. Furthermore, training your cat to accept gentle movement of the skin on the belly in return for reward will help your cat cope with belly palpitation by the vet.

Once your cat is fully comfortable with such handling, it is especially important to make sure that not every instance of the desired behavior is rewarded immediately (see Key Skill No. 8). Not only will this help maintain the relaxed response to handling; it will help encourage your cat to offer this calm behavior when such handling is imperative but no rewards are available, which may sometimes be unavoidable—for example, if your vet has instructed that your cat is to obtain no food prior to surgery, or your cat is too uncomfortable to play as his reward. Furthermore, because veterinary handling is perhaps one of the biggest "asks" we make of our cats, it is really important to give them the signal that shows that this bout of training has finished (see Key Skill No. 9) so the cat knows it no longer has to remain still and calm but can go back to everyday activities.

Remember, none of this training has to be a special event—you can integrate all of this into your daily routine. Do little bits of training here and there, at times when your cat chooses to interact with you. Pick occasions when he is relaxed and happy, willing to engage and the environment is quiet and calm. Once your cat is doing well with you performing this handling, you can ask a family member or a friend that your cat likes to do some of the handling, so he starts to get used to other people touching him in this manner. This will help generalize the association you have created that veterinary-style handling leads to rewards. Do, however, make sure your new handler fully understands the Key Skill No. 5, touch-release-reward—that touch predicts the reward only if the reward follows immediately after the touch has *finished*. In cases where you feel you cannot deliver the reward immediately after the handling, utilize Key Skill No. 4, marking a behavior. For example, with my cats, I remind them of the pairing of the word "good" with treats before I start any handling training, checking that they recall that this specific word means something good is on its way. I can then use the word when my hands are busy with the handling and I need a few extra seconds to get the treats.

Various types of veterinary equipment may be used during examinations, all of them unfamiliar and potentially scary to your cat. Although we may know that a thermometer reads temperature, a stethoscope allows us to hear the heartbeat, and cotton wool can wipe away unwanted fluid, our cats have absolutely no idea of the intention or function of any of these—or indeed that they are being used for their benefit. Thus, having such items come in contact with the body—often in strange places—can not only be perplexing but also potentially distressing for a cat who has not received prior training. Although cotton wool is easily accessible, stethoscopes and thermometers are not common household items. Thus, sometimes in our training we will have to be imaginative in replicating the sensory properties (i.e., sight, sound, smell or feel) of veterinary instruments by utilizing everyday household objects (part of the systematic desensitization process, Key Skill No. 2). For example, a metal spoon can mimic some of the properties of a stethoscope—a round, cool item that can be placed flat against the chest.

As with the handling training, start with your cat on his relaxation blanket in the base of his carrier, so that training commences with your cat in a relaxed state. Then, place the spoon on the floor in front of your cat and allow him to investigate it—hopefully he will start by sniffing it. Reward any positive investigative behavior (Key Skill No. 1). If your cat behaves in a way that suggests he is

Sarah holds Herbie in a standing position with a spoon
placed against his chest, mimicking a stethoscope.

comfortable with the spoon close to him, you can lift the spoon and
touch it gently against his chest for a second, then gently remove it
and reward his calm behavior. Initially, make sure that the object is
at room temperature before you touch your cat with it, because if it
is cold it may be more difficult for him to tolerate it. If he is at all
worried about the spoon, simply spend more time having the spoon
out on the floor, allowing him to habituate to it, and rewarding any
calm behavior directed toward the spoon, even if it is only a glance
in its direction.

For procedures that involve objects placed into the cat's body,
such as a thermometer or a light used to look into the ear canal, it
is never advisable or appropriate to try to recreate such procedures
at home—they should be carried out only by a qualified veterinarian. However, when a cat's temperature is taken, he is often held in

a standing position with tail lifted to insert the thermometer into the rectum. Fortunately, we can at least train our cat to accept standing with tail lifted. Thus, although it is not possible to mimic all procedures at home, we can at least think about what positions our cats may be restrained in and whether it is safe and feasible to teach these at home. Hopefully, you will be able to practice these techniques before your cat ever needs to go the vet.

WHEN YOU ARRIVE AT THE VET'S AND ENTER THE CONSULTATION room, open your cat's carrier door and wait to see whether he is confident enough to leave his carrier voluntarily. If he is, reinforce this behavior with your chosen reward; physical interaction or food are preferable to play at this stage because your cat is about to be handled and we wish to prevent excitement escalating, as it can during play. It is important to reward voluntary actions made by your cat to exit the carrier (Key Skill No. 1). This is not the same as using your rewards as lures (Key Skill No. 3) to encourage him out of the carrier: although such lures may work in getting the cat out of the carrier, they will monopolize your cat's attention while he is exiting the carrier, and he will be too focused on obtaining the lure to take in what is going on around him. Only once the lure has gone, and he finds himself exposed on the examination table, will he take the time to observe his surroundings. If he then finds his surroundings challenging, he will likely shoot straight back into the carrier in a panic. A cat that chooses to exit the carrier at his own pace, assessing his surroundings as he goes and being rewarded for such an action, is more likely to repeat it.

Many cats will not be confident enough to exit the cat carrier, in which case your best option is to ask the vet to examine your cat in the base of his carrier on the blanket associated with relaxation (provided the training for Key Skill No. 6 has been undertaken). That way, the cat will not have to leave his place of safety and he will feel less exposed because of the sides of the carrier. For very anxious cats, placing a towel over them and examining from under the towel, exposing only the areas that need to be examined, one at

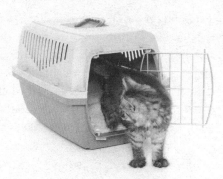

Batman, a kitten trained
by Sarah, voluntarily
leaves his carrier.

Batman receives a food
treat for choosing to
leave his carrier.

a time, can greatly help prevent escalation of stress and reassure the cat that his safety will not be jeopardized.

Tolerating an injection is a life skill every cat should learn. Again, although we cannot actually administer an injection at home, we can teach our cat that the special handling needed is nothing to fear. Generally, injections tend to be given either in the scruff of the neck, or in some hospitals, the back leg or tail. In most cases, the cat is held in a standing or crouching position. When the neck is the target, the loose skin around the back of the neck, known as the scruff, is lifted slightly away from the body with one hand and, depending on the cat, sometimes the head or body is held still with the other. In the case of the leg and the tail, the cat is generally held in the position it feels most comfortable (standing or crouching) and restrained at the front and back end.

Like any other training, the end goal needs to be broken down into small steps, each of which the cat is taught to associate with reward, in a successive manner. For example, the restrained stand or crouch should be practiced on its own, and only when the cat is fully comfortable with this should the gentle lifting of the scruff be added. Only when the cat is calm and content with this should the head or body restraint be added, and finally, an index finger or capped pen can be used to gently push against the area of skin where the injection would be administered, to get the cat used to feeling pressure in this area. It takes only seconds for an injection to occur, but if the cat is scared when it happens, it takes only those few seconds to create a negative association in the cat's mind so strong that he will be anxious on all subsequent vet visits. Thus by practicing these tasks throughout daily life, for just a few minutes at a time, your cat can build up huge amounts of resilience against the actual events when they occur in the veterinary practice.

Sarah lifts Herbie's scruff as a practice for injections.

IF YOU EVER FIND YOURSELF IN A POSITION WHERE YOUR CAT HAS
to visit the vet's in the imminent future, and you do not feel that
you have completed adequate training at home for your cat to cope
with the examination or procedure that needs to be carried out,
there is a "distraction" technique that you can use. This will prevent
the experience from feeling too negative and should prevent the
good work of your initial training from being destroyed. This technique should not replace the training following the format of touch-
release-reward (Key Skill No. 5) in a progressive step-by-step
manner (Key Skill No 2): it is simply a short-term helper for those
unexpected veterinary trips while training is still in progress.

The distraction technique involves providing a continuous supply
of a tasty food to your cat. This can be done by making a runny meat
paste and placing it in a syringe, squeezy tube or small piping bag, allowing you to provide a continuous but small amount of food to your
cat. You can use this as a lure to move your cat into a position (sitting or standing, for example) that you wish him to be in to aid handling (Key Skill No. 3). While your cat is eating the food, you can
touch an area your cat is comfortable with and then gently move to
the new area you wish to touch. The food should be of high enough
value to keep your cat wanting it and therefore staying in the same
position while you touch him. It is harder for your cat to learn that
handling is positive in this situation as the continuous supply of food
is so rewarding that it overrides any association being made in your
cat's mind between handling and food. This is why we do not recommend you rely solely on this method for all your cat's veterinary visits, but just use it in those unexpected and imminent vet visits when
you have not had time to complete the training at home.

To move from distraction technique to rewarding for touch (Key
Skill No. 5), all we need to do is gradually remove the food during
the touch and immediately deliver the food after the touch. This
can be done by holding the food syringe close to your cat but not
pushing the plunger to deliver food until after the touch. Gradually,
you can move to a situation where you present the food-filled syringe only after the touch. This way you will not be using the food

as a simple distraction but will have switched to teaching the cat that handling has positive outcomes and is therefore a worthwhile experience. You can of course also move to other forms of food reward, such as small pieces of cooked chicken or commercial treats.

ENSURING THAT YOUR CAT IS NOT FAZED BY VETERINARY VISITS has many benefits. Such visits are essential for keeping your cat in tip-top health, but your cat cannot know this, and you cannot afford to simply take the attitude that what happens at the vet's stays at the vet's. You may be able to forget the struggle your cat put up while being examined, your vet may be able to dismiss it as one of the expected hazards of the job, but your cat will not forget so easily.

Fortunately, much preparatory training can be carried out at home. Then, by also selecting a veterinary practice that is forward-thinking in minimizing stress on its patients and adheres to feline-friendly handling, you can teach your cat that going to the vet's is as good as a chicken dinner. No matter what health concerns your cat has in the future, he should now be equipped with the capacity to remain calm in the face of new veterinary interventions.

Much of this chapter has focused on training techniques that teach your cat to remain calm and still while veterinary procedures are carried out—such procedures generally involve something touching the cat's body for only a brief period of time. However, there are other circumstances where cats have to cope with things touching their bodies for much longer periods—for example, when wearing a harness. Such a situation also involves teaching your cat not to keep still but to move in a calm manner, despite feeling something strange against his body. In the case of calling your cat in from the outdoors, we wish to train the cat not just to move but to move fast! Such exercises are the focus of the next chapter, which covers the training required for your cat to safely enjoy the outdoors.

CHAPTER 10

Cats at Large
The big outdoors

Is it better for a cat's welfare to live out its life entirely indoors or to have at least some access to the outdoors? This debate continues to rage all over the world. Confining cats permanently indoors is popular in many urban areas of the United States as well as in some parts of Australia. However, in the UK and Continental Europe, particularly in the more rural areas, having access to the outdoors, either via a cat flap or simply being let in and out, is a more common arrangement. Perhaps one of the reasons there is no consensus as to which living arrangement is better for the cat is that each situation has its own sets of benefits and disadvantages, and how these are weighted in an owner's mind depends very much on their individual living situation. If you are one of those owners who thinks that cats deserve a certain amount of freedom to roam where they wish, training your cat can help lower some of the risks that outdoor life entails.[1]

Why might outdoor life be so important to a cat? There are three basic reasons why a cat may feel the need to go out. The first is to seek out the company of the opposite sex for the purpose of mating—this should not apply to the majority of today's pets, which should have been neutered. The second is to go hunting; although many cats seem strongly motivated to try, it appears that only a select few are even slightly successful at hunting. Generally, well-fed pet cats seem to perform in a half-hearted and singularly ineffective manner, and many show little interest in it at all. It's therefore reasonable to ask why today's pet cats should feel any need to go out, period. However, it's self-evident that given the opportunity, many do spend much of their time out in the fresh air. Thus we need to propose a third reason, which is most plausibly an instinctive need to maintain a territory, an area that the cat knows well and regards as his own. Moreover, from the perspective of the cat's welfare, it is self-evident that cats enjoy many features of the outdoors—running at full speed, climbing trees, basking in the sun, rolling in the dust and generally exploring and investigating their surroundings.[2]

Cats are highly inquisitive and exploratory animals, and although it's easy to dismiss such behavior as trivial and inessential, the gathering of information is now known to be a biological imperative, particularly for animals as smart as cats. So far as pet cats are concerned, this may not be entirely essential to their survival, but it would certainly have been important for their wild ancestors, who were specialist hunters and depended for their day-to-day survival on being able to predict where they were most likely to find prey. They would undoubtedly have gathered some of the necessary knowledge directly while they were actually hunting—for example, by remembering places where they'd been particularly successful in the past. However, all the time they would have been soaking up more general information about their hunting area, just by detecting and memorizing relevant clues as they moved around—for example, scent trails left by mice passing by a few hours earlier. Pet cats, even those that don't hunt, undoubtedly feel the same urge to explore and gather information as their forebears did, and although it may not be essential for survival nowadays, it does appear very important for their well-being.

In the days when cats hunted for their living, success would have depended not only on their being able to locate suitable prey but also being the first to find it. For a lone hunter like the wildcat, the greatest competition would have come from other members of its own species. Thus, as well as gathering information about its prey, a wildcat—indeed, any successful predator—needs to glean as much intelligence as it possibly can about the whereabouts of its rivals. In the wild, predators tend to be thinly spread and so rarely meet face to face; rather, they keep track of one another by picking up the other's scent from time to time. During the few thousand years since they have become domesticated, cats have come to live much closer together than their ancestors did, and so can get to know their feline neighbors by sight, but keeping track of them by their odor seems to have remained a priority, judging by how much time cats spend sniffing their surroundings when they are out of doors. You may well have seen your cat rubbing his cheeks against the garden gate or squirting urine in a backward direction against a bush—both examples of how cats leave their mark on their territory.[3]

It's a reasonable assumption that our pet cats still feel the need to keep tabs on what is going on around them, even though for many this is no longer essential to their survival. Only a handful of generations have passed since all cats needed to hunt for their living (i.e., prior to the creation of nutritionally complete commercial cat food), too short a time for evolution to have eliminated the basic survival techniques that permitted successful hunting. Thus, although cats may not understand *why* they feel an urge to patrol the area around their owner's house, nevertheless they still feel compelled to do so, given the opportunity. Once they're out there, many pay only passing attention to the comings and goings of potential prey, because unlike their wild ancestors, they feel comfortably well fed, whatever the time of day or night. Moreover, they still feel motivated to try to keep other cats out of at least some of that area, even though that protectiveness no longer serves any useful function, given that all their most essential resources are within, not outside, their owners' house.

Once they get outdoors, there is always the risk that the cats' territorial nature will lead to fights with other cats and the possibility of injury and infection, particularly if the encounters are with unneutered and unvaccinated males. But the biggest and most widespread risk to cats outdoors is unlikely to be other cats but actually something man-made—*Carrum destructus*, or the motor vehicle. An encounter with a vehicle is often fatal for a cat and thus poses a huge physical welfare risk. Other outdoor risks appear to be country—or even region—specific. For example, in certain parts of the United States, the outdoors puts cats at risk of such predators as coyotes and mountain lions that their European cousins are unlikely to encounter. Risk, however, is often assessed not just toward the cat, but also to other species—by preventing cats from going outdoors, many owners feel they can protect wildlife. Greatest concern appears to be for native songbirds and small mammals.[4]

Risks associated with an indoor-only lifestyle tend to be of a more psychological nature, at least initially. Lack of physical space and physical complexity can cause cats to experience frustration, boredom and even anxiety. For example, open-plan apartments furnished minimally can look beautiful to us, but such vast open spaces can leave cats with nowhere to hide, nowhere to climb and little opportunity for exploration. Living day in, day out, in an environment that does not provide opportunities for cat-specific behavior such as playing/hunting, exploration and solitude can lead a cat to experience chronic stress. Such stress is well known to impact negatively on physical health in terms of weakening the immune system, and thus a psychological risk of indoor-only living can quickly impact negatively on physical health too. Furthermore, cats that have previously been allowed outdoors often take confinement particularly badly, and some may never learn to cope with an exclusively indoor lifestyle, especially if the indoor space is small and uninteresting. The company of another cat is rarely the answer: if two indoor cats do not get on with one another, their opportunities to avoid one another become severely limited. Injury may even result if two incompatible cats are forced to share a home. Confining to indoors several cats who were previously able to get away from one another by going out can produce spectacular levels of tension.[5]

There is, however, a third way of managing cats, little explored so far, and perhaps one that can minimize the risks to welfare that both the other two lifestyle choices bring. This involves allowing the cat outdoors but in a manner that provides a greater level of supervision and monitoring than free access brings (more like the way owners currently manage their dogs). The owner can choose to control the times when her cat is outdoors; for example, she can bring him indoors when it gets dark and he is thus less visible, or when she wants to go out herself and therefore cannot monitor his outdoor activities. Cats can be kept out of trouble by training them to come back on command, just like a (well-behaved) dog. For even more control outdoors, cats can be taught to accompany their owner on walks away from roads and to wear a harness and leash: not all cats like wearing a harness to begin with, but many will do so after a little training.

Cats with restricted outdoor access often appear frustrated at not being able to go out when they want, persistently meowing and pawing at the back door, but they can be trained to be content with this arrangement. Research shows that cats are happiest when they can not only predict what is about to happen but also have a sense of control over their situation, both of which are lost when the cat cannot open the back door for himself or have any idea when it might be opened. Training can give cats the ability to cope with restrictions on their access out of doors, thus reducing frustration and subsequent stress.[6]

Furthermore, the recent and ongoing boom in pet-related technology has provided many devices that aid safer outdoor access. Microchip-activated cat flaps that scan the identification chip implanted between a cat's shoulder blades ensure that only your own cats enter your home, thus allowing your cat access outdoors without the threat of his indoor core territory coming under attack from neighboring cats. Furthermore, some cat flaps have light sensors that allow the flap to lock once it gets dark outside and are even clever enough to allow your cat in if he is still outside, without allowing those already indoors out. A range of specially designed tracking collars and devices that attach to the collar are now available that help locate your cat, some even alerting you if your cat goes outside an area designated by you as safe. Such technology, although appealing to us, will appear utterly alien to your cat, and to maximize its potential, a little training may be required—for example, your cat may need to learn to stand still near the cat flap while the microchip is scanned. He may also need to learn not to be fazed by any lights and sounds the flap makes, and he may need to learn that a slightly heavier collar than normal is nothing untoward.

Thus there are many ways to allow cats to reap some of the benefits of outdoor life, which they still instinctively seek, while lowering the risk of running into outdoor dangers. However, in order to put this lifestyle into place, some preparatory training is required.

I EXPERIENCED FIRSTHAND THE DEVASTATION THAT LOSING A CAT to a road traffic accident can bring and, as a result of this, spent much time putting some of the training tasks described in this chapter into practice.

Cosmos came to live with me with his littermate Bumble when they were kittens. They were great friends and spent a great deal of time outside together, chasing leaves in the wind, dashing up trees and disappearing on adventures farther afield. When Bumble succumbed to a road traffic accident, I was completely devastated—I had never seen either cat venture anywhere near any road. My knee-jerk reaction was to keep Cosmos indoors forever, as the only sure way of preventing such a fate happening to him, and so this is what I did—at least initially. Despite increasing the amount of time I spent playing with Cosmos inside the house and investing in all sorts of cattish entertainment, Cosmos did not hesitate to let me know this was not an acceptable arrangement. After all, he was so used to going outside on a daily basis. For several weeks, he spent most evenings pacing to and fro on the windowsill or in front of the porch door that led to the cat flap, interspersed by frantic bouts of pawing at the door, both interjected with plaintive meows. My heart and my head were in serious conflict. Cosmos wasn't happy, thus I wasn't happy—a strategy had to be sought to help us both be content once more. I spent much time talking with friends and colleagues (vets, behavior specialists, animal trainers and animal welfare scientists) about my options, weighing all the risks and benefits to both Cosmos and me relating to keeping Cosmos inside, letting him outside once more, or even rehoming him.

After considerable thought, I decided the best plan of action for us would be to give Cosmos predictable but restricted access outdoors—allowing him out only at times that I was at home and awake. During this unsupervised outdoor access, I wanted Cosmos to spend as much time as possible in the safety of my garden, but I could not physically restrict him, for example, with cat-proof fencing around the garden, because it is a communal area shared by several residents. By following this plan, I felt I could minimize the risk of road traffic accident or other injury but at the same time let Cosmos have the outdoor enrichment time he so desperately desired. There were many things I needed to do to put this plan into place. First, I needed to make sure outdoor access was predictable for Cosmos so he knew when he would be allowed out and when not, to help him manage his frustrations at not being able to go out whenever he wanted. I also wanted to encourage him to stay close to home, ideally in or close to the garden when outside, and to teach him to come in from the outdoors when required. Unexpectedly, I also managed to inadvertently teach Cosmos to come for walks with me when outside, thus helping me to direct where Cosmos went on ventures out of the garden, without thwarting his freedom. This way, he was free once again to explore his surroundings: sniffing, rolling and even dashing up the odd tree!

Several years on from the death of Bumble, Cosmos enjoys daily outdoor trips. His routine involves going out first thing in the morning. Most days, I no longer need to call him home before I go to work, as he has learned the time at which this happens and on most days, brings himself home through the cat flap just before I am ready to leave. Of course, I reward this with much praise and his daytime ration of biscuits with the odd treat thrown in, distributed across a range of puzzle feeders. After work, Cosmos is again let outside. Because bedtime varies slightly, I do still call Cosmos home at night. He comes home at such a speed I have to make sure to have the door ready and open. It is lovely to see how the training has made the act of running home so rewarding for Cosmos—I think he now enjoys that as much as the food reward or the praise and affection at the end. On weekends, Cosmos has much more time outside as I am at home for more of the day. Cosmos spends a large part of this out-door time in the garden—rolling in the dust, sleeping in the wood-shed, climbing the garden trees and gnawing the catmint.

Prior to Bumble's accident, I rarely saw Cosmos in the garden—he was always farther afield. Spending time with him in the garden playing with twigs and leaves and wand toys, practicing recalls with tasty treats and even playing hide and seek with food treats in the garden has made Cosmos realize that being close to home is fun—it appears he no longer needs his long excursions, when he would be away from home for large parts of the day and evidently ventured far from home judging from the twigs, grass seeds and cobwebs I often found in his fur. Nowadays, he doesn't want to miss out on any of the action closer to home or the opportunity for a treat. The only times when he does leave the garden for any length of time are usu-ally when he voluntarily accompanies me on walks with my little dog Squidge. Although unrestricted outdoor access can never be completely free from the risk of road traffic accidents, I did find a way to minimize the risk to a level where I feel comfortable enough to continue to let Cosmos have some access to the outdoors.

PERHAPS THE FIRST STEP IN PROVIDING OUTDOOR ACCESS IS TO consider whether you would like to use a cat flap. Many owners see no necessity to train their cat to use a cat flap, believing it to be something that they will learn easily of their own accord. This is certainly true for some cats, but learn they certainly have to, be-cause for a cat there is nothing instinctive about pushing its head into an apparently blocked-up hole. The enticement of the outside is often so great that cats are willing to keep pawing and pushing with their heads at the flap until they learn how to move it enough to get out. The reward for such behavior is being outside, and thus if this is something your cat really enjoys, he is likely to do it again and again. However, not all cats learn so easily that the cat flap also moves in the opposite direction, and thus that they can come back into the house through it. Indeed, it may be that what they do learn is that if they just stay outside and meow, you will kindly open the door for them, so avoiding having to learn how to come back in through the flap. For other cats, however, learning through trial and error on their own may take a long time, and some may be too timid to attempt to push their head against a strange piece of plastic. For such cats, and kittens or cats who have no experience of using a cat flap, training can make sure that it is a positive experience and that it can be used both to exit and enter their home.

The best time to start cat flap training is before you have installed it, so that you can start your training completely indoors. If your cat flap is already installed, it may be a good idea to create a makeshift cat flap, just for the training. You could cut a cat flap-sized hole in a sheet of sturdy cardboard and attach a hinged lid from a dustbin or the flap from a litter box that contains a flapped entrance—anything goes, really, as long as it is safe and roughly recreates the situation where your cat has to push a flap to go through a small opening.

For kittens and small or timid cats, pressing their head against the flap can take quite a bit of effort, but there are a few tricks we can use to help. The first is to keep the flap fully open—you can either hold it open, or if you're struggling to hold both treat and flap, you can secure it open with a piece of string taped to the flap. Reward your cat for any investigative behavior (Key Skill No. 1). If your cat will not voluntarily investigate the opening of the cat flap, you can lure him (Key Skill No. 3) by dangling a wand toy in front of or placing food treats around the opening. Once your cat is comfortable eating or retrieving a toy from the opening, it is time to teach him that it is a pleasant experience to actually move himself through the opening. You can toss treats through the opening, hold a treat and move it through the opening, or drag a toy through to encourage him to follow. If you decide to hold a treat and your cat is prone to trying to paw at treats in your hand, it may be best to use the longer stick-type treats or one of the suggested tools to hold the food, such as a syringe filled with a liquid treat. Your cat is unlikely to go through the opening the first time, so remember to reward successive approximations of the final goal; for example, reward for whiskers and nose going over the opening, then only the whole head, then head and one paw lifted, and so forth. Once your cat has gone all the way through, no longer reward the half-hearted attempts, only the entire body through the opening. Make sure you practice with your cat going through the opening from both sides to mimic indoors to outdoors and vice versa.

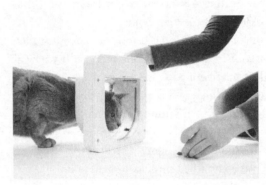

Herbie is being lured through the cat flap opening with the enticement of a food treat.

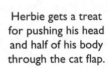

Herbie gets a treat for pushing his head and half of his body through the cat flap.

The next step is to lower the flap and teach your cat to push it with his paw or head, depending on his natural preference. Some cats may find the step from fully open flap to closed flap too difficult, but we can help by holding the flap partially open, either with string or by placing a clothes peg on the flap to open it slightly, making sure we place it on the side that the cat isn't pushing from so that he can still get through. Some cats may still need a lure moved through the flap to entice them through, while others may now go through if they see the lure on the other side.

All cat flaps make some noise—all do as the flap is opened and dropped shut, and, for those that unlock by microchip or magnet, when the locking mechanism opens or closes. Some cats do not mind this at all, but more sensitive cats may find the sounds a little strange at first. For such cats, it is good practice to desensitize them to such noises (using Key Skill No. 2).

Practice both these tasks (getting your cat used to the noise of the flap and going through the flap when it is partially open) several times until your cat is confidently going through the flap. At first, if you are using a food treat to lure your cat through the flap, you may want to place it very close to the flap so that your cat may push his head through and eat the treat and then pull back to remain on the same side he started from. As you gradually move the treat further and further away from the flap, your cat will realize he has to go completely through the opening to get the food treat.

The final step is to remove the peg or your hand holding the flap open, and practice with the cat flap completely closed. Be patient as this is one training exercise in which cats often take their time to move.

If you have practiced all of this with an improvised cat flap in the house, it is now time to install the real thing and practice again with the flap in situ. With time, patience and plenty of rewards, your cat will soon learn to be master of the cat flap.

If you have a microchip- or magnet-operated cat flap, your cat needs to learn to stand for a few seconds while the scanner reads the chip or magnets make contact and unlock the flap. Start your training with the cat flap on unlock mode, and once your cat has fully mastered its use, only then change the settings to locked. Microchip-operated cat flaps can emit a beep once the scanner has successfully read the chip. It is a good idea to desensitize your cat to this noise first (Key Skill No. 2: the noise can usually be created through manual control) so that he does not find it startling. Once he is happy with the noise, it is time to teach him to stay still long enough for the cat flap to unlock.

When I was teaching Cosmos and his brother Bumble to use such a cat flap, Cosmos had a tendency to be impatient and just pawed at the cat flap until it opened. Therefore, I did not have to teach him to stay under the scanner. He soon learned that it was the beep that meant the cat flap would open and not his persistent pawing, and the pawing behavior lessened over time. However, teaching Bumble was a different story. No matter how long Bumble stood under the scanner with me feeding him treats to keep him in position, we never heard the beep that notifies you of a successful scan. It didn't take me long to realize Bumble's microchip had slipped from between his shoulder blades to halfway down his shoulder—completely the wrong position for the overhead scanner to detect! As with all animal training, there are times when you have to get inventive. Because I could not move the position of the scanner (scanners in many newer cat flaps have a wider range and 360-degree scanning), I would have to move the position of the microchip—surgery certainly was not an option for me. Instead, I taught Bumble to lift his front leg up, which rotated his shoulder blade just enough for the scanner to read the chip. How did I do this? I simply placed a sticker on the top corner of the flap, opposite the leg where the chip was resting, and taught Bumble to touch it like you would teach a cat to touch a target stick (see Key Skill No. 3), but with his paw instead of his nose. Training started by placing the sticker on the ground and rewarding any investigation (with nose or paw), then shaping the behavior so that only paw touches were rewarded, and then I gradually moved the sticker off the floor and onto the wall and then onto the cat flap. Because of constant use, it wasn't long before the sticker wore off the flap, but the behavior had been so reinforced— first through food treats and later through being able to get indoors when he wanted—that I didn't need to replace the sticker. Bumble learned that it wasn't the sticker that was important; it was lifting his leg while at the cat flap. The funniest moment was when I had a visitor to the house one day who exclaimed, "I am sure your cat just waved at me through the cat flap"—Bumble had simply lifted his paw to get his chip scanned to come in for his supper!

ONCE A CAT IS ACCUSTOMED TO USING THE CAT FLAP, HE WILL BE fully adept at getting outdoors. The next challenge is to get him back home when you want. Teaching a recall is not just a useful way of achieving this: a recall is as useful for indoor cats as it is for cats with outdoor access. Many cat owners can relate to the mad half hour spent rushing around the house looking under every bed and wardrobe unable to locate the cat, worrying that he has gotten out or got stuck on the wrong side of a cupboard door, only to find him lying as flat as a pancake under the duvet. Teaching your cat to come when called can help to prevent those anxious moments.

For rewards to work outdoors, they need to be really enticing— the outdoors is generally extremely distracting (and often very exciting) for cats—they have space to run and play and opportunities for hunting, as well as exciting smells and sights to explore, not to mention unpredictable noises and the possibility of social encounters. Therefore, your rewards have to be of even greater interest, so that your cat will really think it is worth leaving the wonders of the outdoors to come back to you. Fast-paced games with wand toys and really high-value food treats such as cooked chicken and prawns are often ideal.

I have often heard people say, "I tried to teach him to come but he just turns to look at me then carries on with whatever he was doing!" The problem here is that if you simply call your cat's name, he may not realize you actually want him to come to you. It is likely to get his attention (most cats know their names), but because we voice their names so often, he may not consider hearing his name as an invitation to approach. A new special word or cue is therefore needed to give a clear signal to the cat to approach you. It can be any word you like, but it is likely to be most effective if you do not use that word at other times when interacting with your cat. Words such as "come" or "here" are good cues—short and clear. At the start of your recall training, your cat will not know what this new word means, so you will have to teach him.

Start your training indoors in the place your cat is most likely to pay attention to you. Choose a time when your cat is keen to interact with you—for example, when he is awake and alert, hungry or playful. Place yourself at his level by sitting on the floor at a distance of only a meter or two away from your cat. The closer you are, the less distance your cat has to travel to get to you and thus the greater the chance of success. Once positioned, call your cat's name to get his attention and show him his reward. As long as your reward is enticing enough, your cat should come close to investigate. Once he is beside you, you can present him with his reward.

If your cat does not come, you may need to encourage him by using the luring technique (see Key Skill No. 3). After saying your cat's name to get his attention, stretch out your arm toward him, present the lure (food, toy on string or target stick) and let him investigate it. Then retract it closer to your body. Reward the cat when he reaches you. Once your cat is reliably coming to you, you can introduce your recall cue. Say it after your cat's name and just as he is beginning to move toward you. You may still need the lure at the early stage of this part of the training, but over several repetitions you will find that you will be able to stop the use of the lure, as your cat will have learned that on hearing the cue, coming to you results in a reward. This reward alone should then be enough to encourage your cat to come to you. Remember, you do not have to stick to the same reward every time—in fact, changing the reward introduces an element of surprise that is likely to entice your cat even more to come and find out what is on offer.

Over several practice sessions of just a few minutes, you should find your cat coming over to you whenever he hears the cue word. Depending on his personality and motivation, he may trot over at once, or he may saunter over in his own time. If he always seems to dawdle, you may need to find a more exciting treat. Switching to rewarding intermittently (Key Skill No. 8) will also help speed up the recall. However, the important thing is that he should be coming to you when you invite him to do so. At this stage, you can start to increase the distance between the two of you to several meters. Once your cat is successfully recalling from that distance, you can try moving yourself to another part of the house where he can hear but not see you. Make sure that once you reward your cat for coming to you, you give him the opportunity to leave your vicinity again if he wishes. In this manner, we are teaching him that we just want him to "check in" with us—that coming to us will not always stop the fun he might have been having elsewhere. We do not want him to think coming to you means restraint or restriction, but instead something really tasty or fun. This is particularly important if you are going to build up your recall to be able to use it outdoors.

The next stage is to take what has been learned indoors to the outdoors. Starting very close to the house—in your garden if you have one—is ideal. Just as you began your training in the home at a very close distance to your cat and when he was hungry, affectionate or playful, so do the same outside. As you succeed at each stage of recall training outdoors, change where you call your cat from and the direction you call your cat to, within the garden, including both away from home and toward home. It is really important to use only your super power rewards now that you are outdoors. When your cat comes to you, reward and praise him and then let him explore freely again.

As you develop your recall to coming from the garden to the door of the house, you may start to give your cat the reward on the doorstep, gradually progressing to indoors rather than outside, thus associating the reward with being inside. This way you prevent creating a situation where the cat will come to you but not when you're inside the house.

It is really important to practice recalling your cat at times when you do not need him to come indoors, so that he learns that coming to you does not always mean the end of outdoor time. Stick to this routine both during training and also once the recall behavior is fully established. If the inevitable conclusion to the session is being

Sarah practices Cosmos's recall outdoors.

shut indoors, the association between recall cue and reward could diminish. Thus, interspersing the times you do need your cat to remain indoors with those times where you just want your cat to briefly check in with you will help to keep the behavior reliable and likely to occur again. For many cats, the act of running to you will gradually become rewarding in its own right. Once the recall is fully established, you can vary on which recalls you give a reward and which you don't: this actually keeps your cat more interested in the task, as he never quite knows which recall will bring the treat. However, only progress to this stage once you can reliably get your cat to come whenever you call him. If at any stage your cat stops coming when you call, go back to the initial training stages and reward every recall.

Although training a recall is not a fail-safe way to get your cat to always come home when he is outside, it certainly helps while providing fun and quality time with your cat. Most owners play with their cats indoors but fewer do so outdoors, and by playing these recall games outside, we can encourage our cats to stay closer to home, as they come to realize that it's worthwhile being within earshot in case a reward is about to come.

Now that you are able to let your cat go outside via a cat flap and call him back in when you want, you may decide that the cat flap should be shut at certain times to enhance your cat's safety—for example, you may want to keep your cat indoors at night. If your cat can learn the times when he is allowed outside, he is less likely to spend other times constantly asking to be let out. Not only can persistent attempts to get outside be annoying for you as an owner, such behavior is not ideal for your cat's well-being. Enduring feelings of frustration that the cat cannot alleviate can, over the longer term, lead to a chronically unhappy cat. Luckily, through simple learning processes, cats can very quickly learn a signal that predicts whether outdoor access is available or not. Your first task is to decide when you will let your cat outside and to stick to this routine during the first few training weeks. It may be that you always let your cat out only before and after work, or if you are at home in the daytime, during a time when you are freely available or even in the garden yourself. If you do not have a cat flap and have to physically open a door for your cat, you can call your cat to you (just as you would if calling your cat in from the outdoors) and when your cat approaches you, open the door for him to go outside—this will act as a reward in itself, so you won't need to provide a food treat or a game. If you have a lockable cat flap, call your cat in the same manner to you, but unlock the cat flap only when your cat can see you doing it.[7]

It is really important that you ignore all your cat's attempts to convince you to let him out at other times—this means not only keeping the door shut and locking the cat flap but also completely ignoring any attempts he may make to get outside. These can comprise bursts of pawing at the door or cat flap, incessant meowing or even circling around your feet and rubbing against you while purring furiously. The reason why such behaviors have to be completely ignored (which admittedly is often arduous) is that your cat may perceive the attention you give him at these times as positive, even if you are telling your cat "sorry puss, you are not getting out at the moment." Recall that any attention that the cat perceives as positive, whatever your intentions may be, reinforces any behavior that preceded the attention, thereby making it more likely to happen again.

You can still give your cat plenty of attention when he is not allowed outside, so long as he isn't actually asking to go outside at that precise moment. Do so by entertaining him with other tasks to stimulate both his brain and body and prevent his attention veering back to the outdoors. Such entertainment can involve playing with him, feeding him his meal ration from a puzzle feeder, grooming him, stroking him or undertaking training of another task— whatever he responds to at the time. Soon, your cat will learn that he is allowed out only at specific times determined by you, and as a consequence, he will begin to look to go outside only at these times.

A note of caution: if your cat has previously had free access to the outdoors or sporadic access that could not be predicted, you may find he goes through an initial period where his demands for outside access are even more intense than they were before you began training for this specific task. This is known as a *frustration burst* and is well documented in the training literature. It is simply a heightened reaction to having any desired behavior thwarted. This stage can be difficult for you to endure, but it is really important that you do not give in to your cat's requests at this stage. If you do, all you will have done is to teach him that he has to make a real song and dance before you let him out, creating a situation precisely opposite what you intended. Instead, be strong and persist in letting your cat out at only the predesignated times.[8]

Once your cat has learned the times of day he is allowed out, start to introduce a visual indicator that will tell him when the door will be opened if he asks to go out or when the cat flap is to be unlocked. Having such a signal gives you a little more flexibility in the times you may wish to give your cat outside access, without causing him frustration. The signal needs to be easy for the cat to see, so it is best placed directly on the cat flap or on a door leading to the cat flap at cat height. It could be as simple as a large cross marked on a piece of card—choose a color that creates high contrast to the cat flap or door to make it easily seen by the cat—stuck to the door or cat flap at times when the door is closed. If you always remember to remove this symbol when opening the door or cat flap, your cat will quickly learn that the presence of the symbol means outdoor access is denied and the absence of the symbol means outdoor access is available. Cats are the ultimate control freaks, and by giving them some predictability surrounding outdoor access, we can help to keep them feeling a little more in control and ultimately happy.

Herbie's behavior shows he has learned that the
signal placed on the cat flap means he cannot go out.

ALL CATS THAT HAVE OUTDOOR ACCESS SHOULD IDEALLY WEAR A collar with an ID tag so that it is clear they are owned and not stray, and so that their owner can be contacted if necessary. "Breakaway" collars are recommended because they prevent the cat from ever becoming trapped by their collar, as any tension on the collar will cause it to snap open.

Although most cats cope well with wearing a collar, there are a few that resist it tooth and claw—these tend to be cats that have never worn a collar as a kitten. Luckily, training with rewards using Key Skill No. 2, systematic desensitization and counterconditioning, can tempt even the most resistant cat to wear a collar. Start by placing the collar on the ground, giving your cat the opportunity to investigate it (Key Skill No. 1). Any exploration of it, even a casual sniff, should be rewarded. Then, you can try stretching the collar out on the floor, forming a circle, and placing a treat in the middle of the collar. Hopefully your cat will place his head into the circle

Herbie retrieves a treat from within a collar.

the collar has formed to get the treat. If he does not, it is not a problem—either increase the value of your treat or place it outside of the circle but still close to the collar, and over several training sessions, work toward the goal of having it inside the collar by moving each new treat offered closer to the desired place.

Once your cat is happy to take a treat from within the collar on the ground, you can lift the collar and hold it open on its widest fitting. Cats are highly unlikely to voluntarily place their head through the collar right away (but if they do, do not forget to reward), so we need to provide them with a little encouragement and teach them that such a behavior will result in reward. We can do this by using a lure (Key Skill No. 3), then rewarding the desired behavior. In one hand hold the collar wide open, and with the other hand, move a food reward through the opening, moving it slowly away from your cat, thereby encouraging him to place his head through the collar to obtain the reward. In this particular case, it may be easier to use the lure food as the reward, rather than a different piece of food, as you will have your hands full with the collar and the lure. Initially, your cat is likely to push his head partly or the whole way through the collar without actually touching the collar. Once he has obtained the reward, he will likely withdraw his head. Through the use of a spoon filled with something particularly tasty, such as meaty chunks in gravy or a syringe filled with meaty paste, you can extend the time taken to deliver the reward, thus encouraging your cat to keep his head through the collar.

As your cat learns that keeping his head through the collar results in further reward, you can gently move the collar so that it is resting around his neck, in its final position. The next steps involve repeating this process but with the collar slightly tighter each time, so your cat has to physically push his head through the hole it creates, thereby getting him comfortable with the feel of the collar around his ears and neck. The final stages involve tightening the collar once round the neck to the appropriate amount. The collar should be tight enough to prevent your cat backing out of it or scratching it off, but you should still be able to comfortably get a

A strategically placed treat lures Herbie's head through the collar.

finger under the collar, confirming that it is not overtight. How quickly this training process is completed depends very much on your cat, but most cats, unless they are very timid, will be comfortably wearing their collar within two to three training sessions.

If you wish your cat to wear a tracking collar or device, your cat will have to learn to cope with additional weight and bulk around his neck. One way to teach your cat to accept such an addition is to add a little box to your cat's standard collar, and using the systematic desensitization and counterconditioning principles of Key Skill No. 2, you can then gradually add weight. Some of these locating devices emit a beep when switched on remotely—we can use this to our advantage and train our cats to come to us when they hear the beep, in the same way as we would train a verbal cue such as "come." The benefit of this is, so long as the cat is wearing his collar, he will always hear the cue, no matter how far away from home he is.

ALTHOUGH MANY PET CATS ARE ALLOWED OUTDOORS UNSUPERVISED,
some owners may feel that the outdoor environment holds too many
potential dangers for their cat to have free access outside. In such
cases, supervised access outdoors may be preferable. There are two
ways in which this can be tackled, and which is preferable very
much depends on the individual situation: the temperament and
age of your cat, the type of outdoor environment (urban, suburban,
rural) and the time you have available all need to be taken into consideration. The first way is to train your cat to walk on a harness and
leash, while the second way involves teaching your cat to accompany you on visits to the garden and walks farther afield and not to
stray off on his own adventures.

Without training, very few cats will accept wearing a harness—
cats tend not to like having anything restrictive around their bodies,
and therefore they need to be gently desensitized to such feelings by
pairing the harness with reward (Key Skill No. 2). Harness training is
most successful if it commences when your cat is a kitten, although
some adult cats will learn to walk happily in a harness. Cats of a
bolder temperament seem to fare better (perhaps because they are
better able to cope with the removal of the option to run away when
trouble looms). There are a variety of harnesses on the market—some
have a piece of fabric that tightens around the neck like a collar and
another piece that tightens around the middle of the body, joined together with a piece of fabric running down the back. Other harnesses
appear more like a little sweater that the cat wears with an attachment for a leash on the back. When choosing a harness, the most
important considerations are that your cat cannot slip out of it, that it
does not restrict movement in any way and that it is as comfortable
as possible. The harnesses that consist of a fabric sleeve, rather than
several thin straps, often appear more comfortable for the cat.

Harness training should always begin in the home and at a quiet
time when your cat is relaxed. As with collar training, the harness
should initially be set to a relatively loose fitting and tightened only
once your cat is fully comfortable and confident to walk around wearing it loose. Start your training in exactly the same way as the collar

training, by placing the harness on the floor and giving your cat the opportunity to investigate it, then hold up the part of the harness that your cat's head will go through and lure him through it with rewards. Finally, work toward getting the harness over your cat's head and resting on his neck or shoulders (depending on the type of harness) in the same manner as for the collar. Remember to have brief periods when you are not providing any food reward so that your cat is aware of the harness's presence and associates this with reward.

In these initial stages of harness training, Batman is rewarded with a piece of fish simply for placing his head near the opening.

Batman learns that the sensation of a strip of fabric around his neck results in a tasty reward.

Depending on what type of harness you are using, the next step may be to attach a strap around your cat's belly or lift his front legs through two openings in the harness. Make sure that you move from one step to the next only when the cat is calm and fully comfortable. If the step involves lifting any limb of the cat, be sure to have previously practiced associating this movement with reward (using Key Skill No. 5) before introducing the harness.

If at any time your cat shows any signs that is he is uncomfortable with the harness—for example, trying to back out of it or lowering his body and refusing to stand tall—stop what you are doing immediately, and gently and calmly remove the harness. When you return to harness training, go back a few stages and progress to the next stage only if your cat is calm, relaxed and fully content. Harness training can often take a long time and should consist of many sessions of short duration. Once your cat is comfortable wearing his harness on a relatively loose setting (but not so loose that he could become tangled in it), you can start to gradually tighten the straps so that it fits snugly against his body. Remember to reward generously, as your cat may need a little time to get used to such a sensation—a syringe-fed reward is often useful here to deliver a small amount of food reward frequently. Give your cat plenty of opportunities for positive events such as play or coming for a meal while he is wearing the harness: he should learn that the harness does not restrict movement but instead actually results in something really nice. (As a word of caution, because many harnesses do not have breakaway fastenings, never leave your cat unattended while he is wearing his harness.)

Once your cat will happily wear his harness indoors, it is time to add the leash. Do this indoors and make sure that you keep the leash loose, allowing your cat to choose where he wishes to go. The leash is simply used to keep yourself connected to your cat—it should never be used to pull your cat in any direction. Instead, you can encourage your cat in certain directions with the use of a lure or your recall cue. Likewise, if your cat views the leash as a toy to pounce on, you can use a wand toy to direct his playful behavior onto something more appropriate.

Before introducing your cat to the outdoors in his harness, make sure that he is fully comfortable in it and has had plenty of practice wearing it around the house. The outdoors can be full of unpredictability: some events your cat will take in his stride while others he will instinctively choose to flee from—for example, another cat or a dog. When cats have such encounters when roaming freely, they tend to seek safety by creating some distance between them and the thing they perceive as potentially dangerous, which may end with them hiding under a bush or climbing a tree. A cat on a harness and leash can't do this, which could potentially enhance his concern. A way of getting around this is to provide your cat with a place of perceived safety, which he can utilize when out on walks. A lightweight cat carrier or a pet stroller—which looks like a cross between a fabric carrier on wheels and a child's pram—can be ideal. If one of these is accessible to your cat during walks, he can voluntarily jump in if feeling wary at any point. Likewise, if you detect any potential dangers, you can pop your cat into his carrier, keeping him off the ground and away from danger. Of course, the use of such carriers or strollers does require you to have trained your cat to perceive them as a place of safety before you use it on a walk—this can be done in exactly the same way as cat carrier training.

Your first trips outdoors with your cat on the harness should be short and close to home, and they should occur at a time when the outdoors is as quiet as possible. Make sure that you have a plentiful supply of tasty food treats with you, and a wand toy. You can use both of these as rewards for calm behavior and as distractions from potentially startling situations. Outdoor excursions are not about how much exercise your cat gets, so do not worry if your cat decides to walk only a few paces—all he may simply want to do is explore a small area outside your home. This is absolutely fine, as the opportunity to explore will be enriching in itself. As he gets used to being outdoors in his harness, you will find that he is keen to explore farther afield and for longer periods of time.

SOME CATS NEVER GET FULLY COMFORTABLE WITH WEARING A harness, and as owners we may need to weigh the amount of training it would take to overcome such discomfort against other options that don't involve the harness. For example, my previous cat Horace and my current cat Cosmos received very similar harness training. Horace did not mind wearing a harness in the slightest and was consequently happy to potter around the garden in his harness on a leash. However, although Cosmos would wear his and happily accept treats while doing so, there were subtle differences in his body language in comparison to Horace's that told me he would prefer not to. Further harness training has not changed this dramatically. Therefore, after often having to take Cosmos home after only partially successful attempts to come walking with me and my dog Squidge, I took the decision to formally train Cosmos to walk alongside me without a harness when outside. This has turned out to be a better solution for Cosmos personally. Cosmos and Horace highlight the individuality of cats: sometimes we have to adapt our training to suit their individual idiosyncrasies, as well as the environment we have to work within. Luckily, for Cosmos and me, there are a couple of relatively safe walks we can do from the garden that not only avoid crossing any roads but are also in the opposite direction to any main roads (forming a loop around a residential area with no vehicle through road) and contain a number of neighbors' gardens close to the path. When we do encounter a passerby walking her dog, Cosmos tends to go into the bushes, reappearing once the dog has passed. Because Cosmos knows there are places of safety he can utilize, I do not need to carry a portable safe place such as a carrier or stroller with me. However, if we were ever to extend our walks to other places, these are things I would consider introducing.

Teaching Cosmos to walk with me really just evolved as an extension of our initial recall training. After he would reliably recall to me from different places within the garden and from outside the garden back into the garden, I began to chain lots of short recalls together. After I had given Cosmos the cue to recall to me— "come"—I would move a few steps in the direction away from

Cosmos, thereby encouraging him to follow me. We practiced this over many sessions in the garden, varying the rewards Cosmos got for coming. Soon, I started to notice that Cosmos was voluntarily following me when I was outside—for example, when going to and fro from the house to the car when unpacking the shopping.

I then begun to extend his recall training to coming to me while taking little walks of only a few minutes. I found filling one of my dog treat pouches with cat treats very useful, so that I had treats accessible to reward Cosmos's recalls. Of course, I had to reduce the amount of food Cosmos got in his meal ration to prevent him from getting fat. If at any time Cosmos stopped following me, I would turn and go home (with Cosmos in tow) so that he did not fatigue or find following no longer rewarding, and then later take Squidge for a longer walk without him. After lots of practice, Cosmos was soon following me the whole way on our walks. This is now something I do several times a week. Cosmos will meow to let me know if I am walking too fast. Our walks are therefore a little dictated by him, but I do not mind—it is a pleasure to see him enjoying the outdoors with me, knowing he is not straying or getting into trouble. Squidge also does not mind, as it means she gets more walks and often manages to pinch a fallen cat treat.

THUS, FOR THOSE WHO MAKE THE DECISION TO GIVE THEIR CAT some outdoor access, taking a little time to train a few key behaviors will go a long way to reducing the associated risks. Whether you choose walking on a harness, teaching your cat to use a cat flap that shuts at nighttime, wearing a tracking device or accompanying you on walks depends very much on your own individual circumstances— each should, however, help to reduce the risks in its own different way. Such trained behavior may also be useful in other scenarios— for example, a cat that is able to walk on a harness and leash may suit those that wish to show their cats competitively or take their cats on holidays to places unfamiliar to the cat. Furthermore, a recall is always going to be useful in any situations where a cat may have inadvertently escaped from its carrier or your clutches. A cat that has learned to wear strange things round his neck will cope much better if he ever has to wear a protective collar to prevent him licking a wound after an operation—the foundation training to wearing a slightly more unusual collar is already there, making subsequent training quicker and easier.

For those owners who make the decision to let their cat outdoors unrestrained, there is one consequence they may regularly encounter that they are not so keen on. These are the results of the cat's hunting excursions—that is, dead and injured prey, brought into or close to the home. The next chapter deals with how a cat can be trained to partake in alternatives to hunting that are likely to still meet his needs to perform hunting behavior, albeit in more acceptable ways.

CHAPTER 11

Shredded Drapes
and Bloody Corpses
The less appealing sides of cat behavior

C ATS ARE OFTEN SEEN AS EASY PETS TO KEEP—ANIMALS THAT don't need much attention paid to them, but there when you need a bit of feline attention. However, as we've seen, this is simply not the case. Cats have complex psychological needs. Fortunately for them, there is increasing global interest in animal welfare, and in many developed countries legislation exists to protect the welfare of animals, including pet cats. The laws that relate to domestic cats vary by local government, but in most areas the person who owns a cat is stated to have a duty of care to it. For example, UK legislation makes owners accountable not just for their cats' physical welfare but also their psychological welfare, stating that they have a duty of care to meet their animal's need to exhibit normal behavior patterns. Thus responsible owners and carers of cats need not only a solid understanding of what a cat is, but also the know-how to optimize its well-being.[1]

Somewhat problematically for owners, the pet cat with whom you share your home has in many respects changed very little from its unowned feral counterparts who actively defend large territories (primarily by scent-marking their boundaries) and hunt for their food. When such behaviors occur within the home, the result is often despair, as furniture is shredded by cat claws, or disgust, as another bloody corpse is laid on the kitchen floor. However, as responsible cat owners, it is our duty to prevent our cats from becoming frustrated by providing harmless outlets for such instinctive behavior. With a little training we can teach our cats to redirect such behaviors to targets considered more appropriate, while at the same time maximizing their well-being.

As we've seen, dogs are creatures of habit; cats are creatures of place. Nothing is more important to cats than their territories and the familiarity they bring. Ten thousand years of domestication has done little to diminish cats' overwhelming priority to maintain their own territory, because until recently most cats depended on it for their very survival—it was a place where they could hunt, mate, rear their young and seek shelter and safety. Even today, free-ranging cats still actively defend their territories from intruders, but the sheer distance they often live apart makes it difficult for any cat to warn off all intruders face to face. Instead, the cat uses its complex signaling repertoire to warn others of its presence. Prominent points along territory boundaries are often sprayed with urine, and although such behavior is more common in unneutered males, adult females do spray. The odor of sprayed urine is pungent, and the strength of the pungency is thought to be correlated to the time of deposition—degraded urine actually smells stronger than freshly sprayed urine. This might explain why owners often begin to notice the "tom-cat" type odor in their home only some time after the spraying event. For males, the pungency of the urine is also thought to relate to the fitness of the individual, and thus is likely to be used to attract females as well as repel male competitors. If neutered before sexual maturity at six months of age, most cats will not urine spray, at least not indoors—that is unless they are experiencing profound levels of distress or perhaps experiencing a medical problem. Other chemical compounds are also produced by the cat and deposited within his territory—such chemicals come from the facial glands and interdigital (between the toes) glands and are deposited by facial rubbing and scratching. Facial marking is rarely reported as a problem by owners, because at the very most, repeated rubbing of the same area may result in a sticky brown residue.[2]

However, scratching is commonly reported as a major downside to cat ownership and is even considered a behavior problem by some, despite it being a completely natural behavior. It can be frustrating to discover that although you've provided your cat with a pristine scratching post, he chooses instead to scratch the stair carpet, the sofa, the wooden banisters, even the wooden door surrounds. For occupiers of rented property the repercussions can be costly. Even among professionals, opinions are strongly divided as to the best solution. Surgical declawing is practiced in some parts of the United States but is illegal in the UK and much of Europe. We are firmly in favor of behavioral methods for addressing scratching in unwanted places.[3]

Scratching is instinctive and therefore would cause frustration in the cat if prevented: it is also habitual and therefore can be difficult to redirect. It is undoubtedly also a pleasurable behavior for cats— you may even witness your cat becoming playful or having an excitable dart around after he's finished scratching. Cats tend to use the same scratching site over and over again, which does result in a clear visual sign that reminds the cat where he has been scratching (as well as in the olfactory signal, which the human nose cannot detect). Furthermore, scratching sites in free-ranging cats are distributed along regularly used routes rather than at the periphery of the territory, and therefore even if a cat has outdoor access, he is likely to continue to scratch inside the house, possibly at multiple sites. Scratching also appears to fulfil other functions beyond marking territory—it helps keep claws in optimal condition and often forms part of the stretch after resting.

As well as clearly marking it, cats also use their territory (and even land beyond) to hunt. Not so long ago, cats were praised for their hunting prowess. Now, in the course of just a few decades, cats have come to be demonized as "murderers," and even their doting owners can be appalled when they bring home gory little "presents." When wildlife populations decline, especially in habitats adjacent to housing, cats are regularly blamed as the most obvious culprits; the cat's habit of bringing its prey home (and then often leaving it rather than eating it, because commercial cat food is tastier) does it no favors at all. Predation by unowned feral cats can undoubtedly wreak havoc in some habitats, especially those that are isolated, such as oceanic islands where there are no competing predators. By contrast, the impact of pet cats on mainland wildlife has been much harder to prove, and most conservationists now agree that habitat destruction and fragmentation pose much greater threats than a few pet cats ever could. Nevertheless, this cannot hide the fact that pet cats do kill many millions of birds and small mammals each year in the UK and United States alone, and many more that do not kill are likely to spend part of their day trying.[4]

So why is hunting so important to the well-fed pampered pet? The first reason is the cat's addiction to animal flesh. Cats obtain most of their energy from protein and fat, and not from carbohydrates as we do: a starving cat will literally be digesting its own muscles, unable to stop its body from breaking them down to provide fuel. Moreover, such a cat needs not just any old protein but specifically animal protein, as protein from plants lacks some components essential for cats' health, components that dogs (and humans) can make inside their own bodies. Furthermore, female cats need to take in a certain amount of animal fat, the only source they can use to make the hormones that regulate their reproductive cycle—no meat, no kittens.

The consequence of this addiction to meat is that prior to the introduction of commercial cat food, only those cats that were the most skilled hunters were able to survive the lean times when prey was hard to come by. Just twenty-five generations back from today's pet cats—cats that we would dearly love to desist from hunting— only the leanest and meanest hunters would have left enough offspring to qualify as ancestors. It was less than fifty years ago that nutritionists were able to work out all of the cat's nutritional peculiarities and design commercial cat foods that were guaranteed to be nutritionally complete. That's why almost all cat foods you buy in the supermarket contain a substantial proportion of meat or fish.

So despite there nowadays being ample supplies of animal protein available to cats in an array of meat-based diets designed specifically for them, cats are ultimately descendants of the most proficient hunters. Too little time has passed, and with no selection bias from humans for those that are less able at hunting, for the instinct to hunt to subside.

When Thomas Harris, author of *Silence of the Lambs* and *Hannibal*, wrote, "Problem solving is hunting. It is savage pleasure and we are born with it," he was describing humans, but these words also perfectly capture two important features of a cat's experience of hunting. The first is problem solving—hunting is not easy; it takes time, dexterity, physical exertion, cognitive skill and requires acute sensory concentration. Also, a great deal of learning occurs during every hunt—one false move and the prey is lost. In fact, the average pet cat generally has many more failed attempts at hunting than successes. Mice can disappear in and out of vegetation, rabbits can scoot down holes and birds fly off if approached too quickly: thus each type of prey presents a cat with a unique set of problems. Second, hunting is pleasurable—the reward center of the brain releases endorphins as a cat pounces or bites a prey item. This, coupled with the reward of the opportunity to consume flesh, makes hunting a well-reinforced behavior.

Even the most laid-back pet displays its legacy as a hunter when it plays with "toys": this may be how we refer to them, but cats regard them more seriously than the word implies. Dangling a toy mouse in front of your cat can provide hours of harmless fun for both parties. Crucially for our understanding of how cats' minds work, the cat generally appears more interested in the toy than the human on the other end of the string (in contrast to dogs, who mainly use toys as a way of interacting with people). In fact, scientific research has revealed that cats do treat toys just like they do prey. They mostly prefer toys that are about the size of a mouse or a small bird, they like their toys to have limbs and either fur or feathers, and they play with large toys at paws' length as if they were worried they might bite back. The toy needs to move or fall apart if it is to sustain their interest for long, and, most intriguing of all, they play more intensely when they are getting hungry. All of these traits mirror exactly what we know of what motivates cats when they're hunting, and so it seems likely that when they are playing, they think that the "toy" is actually prey.

This raises the interesting possibility that playing with toys could be an effective way of meeting the cat's desire to hunt, both for an indoor-only cat and even for a cat that does go out, a way of using up energy that might otherwise go into actual hunting. Thus, by teaching our cats that we provide ample opportunities for hunting behavior, we may just be able to satisfy their predatory needs without any blood being spilled.

So we understand that cats need to mark their territory (even in the absence of competitors) in order to feel secure—it's just a cat thing. We also understand that they need to hunt, or at least feel as if they are hunting. So, how do we set about training cats to perform such behaviors in ways considered acceptable by both owner and cat?

The first step in teaching your cat to scratch in acceptable places is to provide suitable objects to scratch on. Very often commercial scratching posts are too short or if tall, have platforms incorporated that block the cat's access: a cat needs to be able to stretch to full height when scratching. A cat that attempts to scratch on these will quickly learn it cannot fully stretch and on finding the action of scratching uncomfortable or unfulfilling, will seek out easier options. Therefore, look for a scratching post that is taller than your cat's length when he's stretched out. We have witnessed many cats who have access to such scratching posts literally jump onto them with all four feet and scratch both their front and back claws in apparent ecstasy. Commercial scratching posts tend to be made of sisal rope, but you can also make your own using wood, carpet or sisal textile material glued to wood (cats often prefer the coarser backing of the carpet to the soft pile). Unfortunately, not all cats scratch only when standing upright; some like to scratch horizontally too, and any freshly laid carpet in your new home may be particularly tempting. Corrugated cardboard cat scratching pads are available commercially, but you can also make horizontal scratchers by firmly attaching your cat's preferred scratching material to a block of wood. Stability is a really important factor—any scratcher that wobbles midscratch may startle your cat, and he will soon learn to avoid that location. Make sure that the scratcher has a large and heavy enough base or is secured to the floor or wall so that it offers plenty of resistance when your cat pulls its claws against it.

Second, the scratching surfaces need to be positioned in the most appealing places to scratch within the home. If your cat has already shown signs of scratching somewhere other than the structures you have provided, it is a good idea to position additional scratching places at the location of the inappropriate scratching. Provide scratching sites near to doors that lead to the outside (possibly perceived as territorial boundaries by your cat) but also along regular routes he takes through your home. For homes with stairs, owners often complain that the banisters or staircases are common targets—this is likely a route that is well traveled by their cats. If your cat is one that tends to scratch when he wakes up, it is a good idea to have a scratching place near his favorite sleeping place as well.

For many cats, simply having scratching resources in the right locations is enough for them to use them spontaneously—and to make their mark in the home without damaging it. However, other cats need a little formal training to learn where you want them to scratch. If your cat appears to be one of the latter, you can encourage him to interact with his scratcher by luring him there (Key Skill No. 3). An ideal lure for this training task is a toy on a wand. Move it quickly around the scratcher, pulling it slightly out of reach to tantalize your cat so that as he enthusiastically jumps and swipes for it, he makes contact with the scratcher. His excitement, paired with the feeling of claw touching the scratching material as he reaches for the toy, is highly likely to set off a bout of scratching. If it does, let him know you are happy about this with lots of praise and further rewards of your choice. A verbal marker that a reward is coming (Key Skill No. 4) is ideal in this situation, as you do not want your cat to break off from scratching to receive his reward; rather, you want him to know that a reward is on its way as soon as he has finished scratching.

If he doesn't scratch, make sure you still reward his behavior of following the lure. After all, the more he learns that being near the scratcher brings reward, the more time he will spend there and the greater the chance he will start to use it to scratch. Keep practicing with your lure every few days until you regularly see scratching at this site. You can further encourage your cat's interest in the scratching site by collecting the scent from his facial glands (Key Skill No. 7) and rubbing it on the scratching site. For cats that are attracted by catnip—not all are, as this response is determined by the cat's genes—you can also sprinkle some dried catnip around the scratcher. (Cats that are excited by catnip will usually roll over and then often kick and claw objects nearby.)[5]

SCRATCHING IS A NUISANCE TO OWNERS, BUT HUNTING ALSO HAS repercussions for the local wildlife as well as attracting the disapproval of the authorities in a few parts of the world. Luckily, with the correct techniques you should be able to teach your cat to prefer the outlets for hunting behavior that you provide over the real thing.

Predation involves a whole sequence of behavioral steps, starting with locating the prey, followed by capturing (stalking, chasing, pouncing), killing, preparing and eating it. Because some of these actions are hardwired into every cat, it is unlikely that training could ever "switch off" predation completely. However, training can be used to channel the building blocks of hunting behavior toward acceptable targets—"toys"—thereby providing predatory opportunities at the same time as creating a "game" that is rewarding for both cat and owner.

Some cats inadvertently train themselves that an altogether different target can be a great way of venting predatory behavior—that is, they "hunt" their owner's hands and feet. This undoubtedly has painful consequences for the owner, and as a result, can lead to a relationship breakdown where the owner begins to fear the cat. Fortunately, training can be used to redirect this behavior toward toys and other more appropriate "prey."

Our goal is therefore to reduce or discourage hunting on live prey (and human hands and feet) by providing the cat with games that are as rewarding, cognitively engaging, time consuming and physically tiring as real hunting. If we can fill the time a cat would otherwise allocate to hunting with hunting games that do not involve real prey, we are likely to satisfy the cat's inclination to hunt not just physically but also, and perhaps more importantly, mentally. Even if his instincts kick in when he catches a glimpse of a fluttering bird or scuttling mouse, our hope is that the cat will be too unmotivated to hunt properly, leaving the prey free to escape. In addition, incorporating hunting-style games into your daily routines with your cat has the advantage that he will be much less likely to become overweight, bored or frustrated—as can happen to cats who have little else to fill their time with. In some countries, allowing a pet to become obese is classed as one form of neglect, in the same way that allowing an animal to starve is another. Thus, teaching your cat to engage in appropriate hunting alternatives will help promote your cat's welfare in not one but several ways.[6]

A cat with no other supply of food needs at least ten successful hunting trips daily, each resulting in a catch of a single mouse, to meet his daily nutritional requirements. However, he will need many more hunting attempts than this to meet these requirements, because many will end in failure. Thus, although the attempts to capture prey may be rewarding in themselves, the jackpot reward of catch, kill and consume inevitably occurs intermittently. Recall that once a behavior is learned, it is most likely to be maintained if it is intermittently reinforced (Key Skill No. 8)—that is, when the

cat does not receive a reward every single time it performs the behavior. Creating such intermittent reinforcement in hunting games
should therefore strengthen the rewarding nature of such games for
your cat. Therefore, when playing with your cat, make sure that he
doesn't end up the "winner" every time. For example, if the game
involves a wand toy with some feathers on the end, make sure that
some pounces or swipes at the toy do not result in the cat obtaining
the feather toy. If you've laid out a series of hidden toys or treats,
arrange it so that not every hiding place you use contains a reward
each time you lay out the treats.

Because they are often unsuccessful at obtaining prey, cats without alternative sources of food have to spend large parts of their day
in hunting excursions. However, much of their time is taken up
traveling, often across great distances, to known hunting sites, navigating the complex environment on the way. Thus, the energy ex-
pended before a prey item is even in sight is often great and is the
main way a free-ranging cat keeps its athletic physique.

Many of the games owners play with their cats focus entirely on
the capture part of the predatory sequence, omitting all the steps
that usually lead up to this. Couple this with a diet containing more
calories than the cat needs, and very soon a cat can become overweight and lazy, no longer interested in the games he used to enjoy.
There is a way, however, that we can provide our cats, even those
kept exclusively indoors, with an outlet for the environmental navi-
gation that hunting for prey entails.

For many pet owners, the word "agility" conjures up thoughts
of a handler guiding a dog through tunnels, over jumps, along walkways and weaving through poles in a competition. However, agility
needn't be reserved for dogs, nor does it need to be competitive. It is
now possible to purchase scaled-down outdoor agility equipment for
cats, but it is just as easy to make your own, using branches as jumps
and open-ended cardboard boxes as tunnels. Bamboo canes pressed
into the grass make ideal weaving poles, and any garden furniture
can be used for training your cat to jump up and down. The purpose

of agility with your cat is not to test how obedient he can be or how well he can learn commands (although you can of course teach verbal cues to jump, weave and go through a tunnel if you so desire). Nor is it about how fast your cat goes around the course. It simply provides an opportunity for your cat to engage physically with its environment using behaviors it would use when searching for prey—for example, jumping over fences or fallen trees and navigating through undergrowth. As well as stimulating both your cat's mind and body in a way that it would be during predatory behavior, teaching your cat agility and hunting games is good fun and can greatly enhance the bond between you and your cat.

Before you start, give your cat time to explore any new equipment (Key Skill No. 1). Once your cat has shown that he is comfortable with his new agility setup, it is time to select your lures (Key Skill No. 3)—wand toys on fixed rods or wire-based wands make ideal lures for your cat to follow over jumps and through weave poles. You can also toss treats through tunnels to encourage your cat to follow them and thereby learn to use the tunnel. As your cat becomes acquainted with the equipment and comes to associate it with rewards, you will likely notice an increase in your cat's speed and skills. Make sure your agility sessions are short, particularly if your cat is overweight or unfit, as they can be extremely energy demanding and cats do not find it as easy to get rid of heat as dogs do.

Agility, games and other training exercises do not have to be reserved for outdoors: on the contrary, it is actually very beneficial for exclusively indoor cats, as it provides them with an opportunity for high-energy behavior and exercise that they may not normally have indoors. Within your house, think how you could organize your furniture and belongings to encourage your cat to become more active. Are there pieces of furniture that are close enough together or can be moved closer together so that your cat could jump between them? Do you have, or can you get, some cardboard boxes that your cat can jump in and out of or run around or through? Do you have any shelves your cat can walk along? Think about which behaviors your

Luring Herbie over a jump.

Herbie learning to follow a
target stick through weave poles.

cat might use if he were navigating the outdoors on a hunting excursion.

Rethink your home as a potential cat hunting ground. Many owners prefer their cats to keep their four paws firmly on the floor and off the furniture. However, cats were designed to climb and jump, and when left to their own devices, they explore and utilize vertical space just as much as horizontal space. Instead of discouraging this side of your cat's nature, provide furniture and shelves that your cat is allowed to use. By encouraging climbing, jumping and generally being off the floor, not only will you provide a range of vantage points, you will help develop your cat's balance and coordination.

Begin with two pieces of furniture almost touching so that your cat simply has to walk from one to the other to obtain a reward. Gradually move them apart so that your cat has to step across the gap, building up to a gap big enough that your cat has to jump across. By moving objects in this way, you are progressively building up the amount of exercise and mental effort your cat has to exert to reach his reward. To encourage your cat to move from one piece of furniture to the other in the early stages of training, you can use a lure (Key Skill No. 3).

Agility is one of the areas of training where I find a verbal marker (Key Skill No. 4) very useful. This is because the behavior tends to occur quite fast and I want to be able to let the cat know what exact behavior he is receiving the reward for. For example, I cannot deliver a treat while the cat is in midair, but I can say "good" at this time, and my cats have previously learned that the word "good" means a food reward will soon follow.

Once your cat has mastered the jump between two pieces of furniture, you can start to add other activities, such as jumping down to the floor and back up to the furniture, or going around a piece of furniture. Teach each individual task separately, and once your cat has them mastered, you can string them together to create indoor agility courses.

The agility course needs to be designed with your cat's individual requirements in mind. For cats that are overweight, make sure that

Herbie learns balance during indoor agility.

your chosen tasks involve minimal physical activity, to allow your cat to build up fitness. Place jumping objects very close together, and for kittens or elderly cats, position the objects close to the ground and make sure they are stable. If you have a cat with mobility problems such as arthritis, which make jumping painful or difficult, stick to less physical tasks such as simply walking from one piece of furniture to another, rather than jumping. For cats with any medical conditions, please consult your veterinarian before undertaking any agility training, either indoors or outdoors.

Just as agility training provides a wonderful opportunity for your cat to utilize his environment in a more physical way as he would when on a hunting excursion, hunting games can provide an outlet for the predatory sequence that occurs after location: capture, kill, manipulation and consumption. Once a cat has located a prey item in its vicinity, the actual time a cat spends exerting energy in catching this single prey item is rather brief—a single pounce takes only seconds—and if the cat is not successful with his first pounce, he is unlikely to be offered another chance by the same prey item, which will have fled as quickly as possible. Consequently, hunting games can be relatively short, but they should be frequent—you will have more success at engaging your cat in hunting games if you intersperse short games throughout the day as opposed to dedicating a single block of time for all of the day's hunting activity. This may be difficult to achieve if you are not home for much of the day. Luckily, as well as games that require you to be a player, there are those that can be set up that do not require human interaction—games that can entertain your cat while you are out or busy.

Pet cats are generally opportunistic hunters: it is often the sight of a potential prey item darting past in their peripheral vision that brings their hunting instincts alive. We can utilize this in our hunting games, enticing the cat to play by moving toys quickly past him in his peripheral view. Cats struggle to focus on static items that are very close to them. Therefore, position the toy at least a cat's length away and ideally to one side of him. You can then gain your cat's interest in the toy by moving it in the same way a prey item would move. If the toy is small and furry like a mouse, it is best to move it along the ground, away from the cat in fast, straight lines, mimicking the way a mouse scuttles.

I have yet to see a mouse run toward a cat, but I do frequently see owners use a wand to whip a toy *toward* their cat. In these circumstances, some cats appear to fixate on the end of the wand rather than the toy—this is usually because the wand is moving in a quick and straight manner while the toy at the end of it is darting all over the place. This is generally the case for wand toys where the toy is attached to a long piece of elastic. Wand toys that incorporate a fixed cord tend to work better, particularly if you are moving the toy from a seated or standing position. Alternatively, you can move around dragging the wand toy behind you—keeping you fit as well. Often commercial wand toys have short wands. If you have a cat that gets excited during play and you are worried about him accidentally catching your hand with a claw or tooth, you can attach the toy to a longer homemade wand such as a bamboo cane or to a horse's schooling whip or lunging whip for a really long wand useful for dragging.

To really get your cat moving, you can attach one of his toys to a child's toy fishing rod and cast the toy out and reel it back in. Plastic fishing line that will not hurt the cat's mouth if he tries to bite it is best, as generally the cat does not see it, giving him the impression the toy really is moving on its own. Fishing rod games are brilliant for outdoors or down a long corridor where there is ample space to cast the toy several meters.

During each game, cast the toy no more than a handful of times. This will ensure that the game stops when your cat is still finding it rewarding, making him eager to play the next time he sees the fishing rod. It also prevents your cat from becoming overly fatigued. Cats do not chase their prey to the point of exhaustion—they are sprinters rather than marathon runners and often prefer a sit-and-wait strategy with a short burst of energy at time of capture. Thus, although some chasing is good fun and energy expending, their bodies are not designed for long chases at high speed.

To mimic airborne prey, choose a wand toy that is light in weight; those made of feathers are ideal. Such toys tend to be attached to a fixed wand or a wire and are better moved through the air than across the ground—swoop the wand through the air in a curve, perhaps bringing it to rest momentarily on the floor before gliding it back through the air. This will encourage your cat to spring upward, jumping to catch the toy between his front paws. Because such toys elicit the stalk and chase part of the hunting sequence, which is innately rewarding in its own right, there is no

Sarah casts a fishing rod toy for Herbie.

need to reward such play behavior with additional treats such as food. In fact, during this stage of hunting, the cat's senses are so focused on capturing the toy in front of him that eating will not be on his agenda.

How frequently you should allow your cat to actually capture the toy should be determined by how engaged he is in the game. If he is not so confident or not fully engaged, allow him to capture it more often than not. If he is fully engaged and adept at catching the toy, make it slightly harder for him to catch, challenging him slightly more and ensuring that there are times when he simply cannot catch it. In this way we can exercise the cat's problem solving abilities, making him think about how he can capture the toy next time. You may have already experienced the scenario in which a cat who cannot reach a wand toy jumps onto a nearby piece of furniture in attempts to have a better chance of reaching it.

Regardless of the number of failed attempts to catch the toy, as you bring the game to an end, start to move the toy more slowly, mimicking a prey item that is getting tired or that is injured, encouraging your cat to naturally slow down his movements too and giving him every opportunity for one final capture. Once your cat has completed the capture, he is given plenty of time to kick, bite and hold the toy. Only once he loses interest in it should you remove it and put it away out of sight but reach-ready (ideally, in your training toolbox) for future games.

One toy that can never be truly caught is the red or green dot produced by a laser pointer, which is readily available commercially as a cat toy. The laser creates a small dot of light when directed toward physical objects and can easily be moved quickly and in straight lines. As a consequence, the laser pointers can very quickly "switch on" a cat's hunting instinct. However, as the light point can never truly be captured, such a situation may leave a cat feeling frustrated. Moreover, some cats have been known to become fixated on other sources of light, such as the reflection from a watch in sunlight, after playing with laser pointers. Laser pointers should therefore never take the place of a wand toy. If they are used, they should be in conjunction with toys that a cat can physically manipulate. For example, you could use the light to lead the cat to an actual toy on the floor, then switch off the light as he pounces on the toy: however, not all cats are fooled by this and will continue to search for the elusive light.

Most cats like to chase toys but a select few also like to retrieve, often dropping the toy at their owner's feet in attempts to encourage their owner to throw it again. For such cats, the chase and fetch sequence is rewarding enough for the cat to perform it over and over again. It's puzzling why some cats seem to enjoy this dog-like behavior so much. However, mother cats will bring prey home for their kittens, and adult pet cats often carry their prey home, presumably to eat it in a place they consider to be safe. If you see your cat carrying any of his toys in his mouth, you can "capture" this behavior by throwing the toy as soon as he drops it. It does not matter if he does not bring the toy all the way back to you—what is important is that he shows that he enjoys the chase and the carry. If you have already taught your cat a verbal marker as a signal that a reward is on its way, you can also reward this behavior with your verbal marker and follow up with your chosen reward when your cat reaches you. It is never a good idea to try to forcibly remove the toy from your cat's mouth: simply wait until he chooses to drop it. If the cat considers the toy to be prey, a human hand trying to remove it from his mouth is only likely to make him clamp his teeth around the toy so as not to lose it.

Cats that catch birds need to pluck away some of the feathers before they can consume the flesh and innards. Then, as they move to eat different parts of the bird, more feathers may need to be plucked. For a few cats, such hardwired behavior can be directed toward other items, often annoying the owner. Examples include plucking the paper from toilet rolls and biting off small pieces of cardboard from boxes. Some cats even resort to plucking fabric—in fact, a common part of the treatment plan for the clinical behavioral problem known as pica, where the cat ingests inedible items, is providing the cat with something it can pluck. Thus, for cats that like to pluck, providing an appropriate outlet may help unwanted tearing of other items while fulfilling his need to pluck. Even if you have never seen your cat attempting to pluck anything, it is worth providing your cat with a destructible toy such as a cardboard roll to see if it is a behavior he enjoys.

Some owners worry that by providing a cat with something more appropriate to pluck, they are encouraging the behavior and feel that the animal will pluck more frequently in the future. This should not be the case. It is almost impossible to remove the urge to perform hardwired behavior: although physical prevention may temporarily stop the behavior, it does not remove the necessity the cat feels to carry out the behavior. In fact, the animal may actually be

Cosmos wrestles with a plucking toy.

more likely to perform that behavior, or perform it more intensely, after the physical restraint is removed. By providing a suitable alternative object to pluck, we give the cat an acceptable way to fulfill his needs and thus reduce the likelihood of his performing the behavior elsewhere. Some cats enjoy plucking cardboard—if so, provide them with boxes that they can do this to. Other cats pluck only feathers, in which case you can make holes in an empty cardboard toilet paper or kitchen roll and fill them with feathers—I often collect fallen feathers on walks for later use as cat enrichment. To combine plucking with the subsequent behavior of consuming, you can also fill the cardboard roll with wet or dry cat food, covering the ends of the roll with baking paper pierced with little holes to allow the cat to smell the food inside. For those with creative minds, the sky is the limit—you just need to always avoid any materials that would be dangerous for your cat to ingest. If you are unsure, check with your veterinarian.

When handling larger prey such as rabbits, cats often lie on one side and rake and kick the prey with their back legs. You may have seen your cat do this to a toy—the behavior may start with the cat standing over the toy and repeatedly tapping it with one back foot, almost as if he is trying to start a motorbike. The cat then flops onto one side, pulling the toy round onto his middle with his front paws, while kicking and raking the toy with his back feet. This behavior appears to be reserved for larger toys, and it is therefore important to provide your cat with these as well as smaller, more chaseable, items. You can make large toys yourself by stuffing old socks that you then decorate with feathers. For cats who enjoy catnip, some catnip inserted in the stuffing will provide a fun-filled few minutes.

Generally, cats do not like to get wet, rushing indoors when the rain comes down. Not all, however. You may be all too familiar with the scenario of a keen feline face and two front paws that appear as soon as you relax into a bubble-filled bath, eager to pat your toes glimpsed through the bubbles. Again, the lure of something dipping in and out of view is too much for the cat to resist. You can encourage your cat's "fishing" abilities by providing more appropriate outlets, such as ping-pong balls and small plastic toys bobbing in a children's wading pool or plastic bucket. Adding bubbles will only make the allure greater as the item will then disappear and reappear. A bold cat may even enjoy watching and trying to capture a toy that moves automatically through the water—for example, those with a drawstring or wind up mechanism. One word of caution: make sure that the vessel containing the water cannot be knocked over, creating a flood, and keep the water level low, just in case your cat does decide to jump in. For those with a garden, a wading pool can be a great addition, as fallen leaves and petals will only make the pool more enticing.

Cats' quarry often hide in the undergrowth, trying to conceal themselves from prying feline eyes. Therefore, the slightest rustle or movement from the undergrowth can send a cat into a heightened state, with every sense fixated on the point where he last saw or heard the prey. We can mimic such sensory acuity through hunting games, both inside and outside the home, using nothing more than a toy and a cardboard box. To prepare the box, cut out small holes

Cosmos pond-bobbing.

just wider than your cat's leg, at various places on the sides of the box. Sit on the floor with the opening of the box facing you so the cat cannot get into the box but can see the upturned base and sides. Using a small toy, ideally one that is attached to a fixed wand, push the toy partially through one of the holes to catch your cat's attention. Once you have his attention, you can remove the toy and push it partially through another hole. This way the toy is never fully on view and such a presentation will entice the cat to push his paw through the hole in attempts to locate it. We have in effect created a feline version of the popular arcade game Whac-A-Mole. Toys that emit sounds such as electronic cheeps and squeaks add extra excitement to this game and help encourage the cat to take part in patting the toy. End the game by allowing your cat to capture the toy and, if he so desires, push it through the hole toward him for more fun.

Similar games can be played with a toy on a fixed wand held under a fabric sheet, pillowcase or sheet of newspaper, with the toy being unexpectedly "peeked" out from underneath. Make sure you have a long wand attached to the toy as your cat is likely to pounce

Cosmos and Sarah play Whac-A-Mole.

down on it while it is still partially hidden. Some commercial products are available that have a battery-operated toy hidden in a bag or under a piece of fabric— thus providing similar concealed movement of the toy at times when you are not free to play. A word of caution: always try these toys for the first time when you are there to supervise, to make sure that your cat finds them enticing and not frightening.

The majority of cat owners feed their cats their daily food as two meals from a bowl, usually located in the same place, day in and day out. However, if we think about how cats hunt, we can quickly see that this is very different than how they would obtain their food if they were wild. Wildcats would be eating little and often, exerting much mental and physical energy to obtain their food and finding each catch in a different place, even if within the same hunting patch. Luckily for us, there are simple changes we can make in our routine that will teach our cats that obtaining their food at home can be just as rewarding as hunting. By using their mental and physical energy to obtain the cat food (which ultimately tastes better than any prey they might catch), they may be less likely (and have less time) to hunt.

Puzzle feeders are a great way of prolonging the time taken to access and eat food, and they are an excellent way to keep your cat occupied when you are not around to entertain him. Commercial puzzle feeders come in a host of different shapes and sizes, ranging from balls with holes that can be filled with dried food that falls out as the cat learns to roll the ball along the floor, to plastic mazes that the cat has to pull the food out of with his paw. Homemade puzzle feeders can be made by cutting suitable holes in a cardboard box lid and inserting small pots (ideally of various sizes) into the holes to create a series of tubs in which to place biscuits that your cat can fish out with his paw. Tubs and small boxes and tubes can also be taped or glued directly onto the cardboard to give the cat a variety of heights to work at. Cardboard rolls for toilet tissue or paper-towel rolls can be glued or taped together on their sides to create pyramids that cats can slide their paws into to extract their biscuits. When providing your cats with such items, don't be afraid to move them to different locations within the house so your cat has to explore in order to locate where his food might be hiding.

If you are at home and able to feed your cat more than twice a day, divide his daily food ration into several smaller portions and provide these in the puzzle feeders at various times throughout the day. Scientists have shown that even when provided only twice a day, puzzle feeders extend the bouts of eating and the total time spent obtaining food in comparison to the same weight of food being provided in a bowl. In fact, you can even turn your home into one giant puzzle feeder by hiding little open pots containing a few biscuits or a spoonful of your cat's wet food around your house for your cat to hunt out. If for any reason your cat cannot use a puzzle feeder—for example, due to mobility problems—commercially available timed feeders can be used to disperse the meal rations across the day, helping to mimic the small and frequent meals cats would get from hunting for their food.

In addition to, or as an alternative to, puzzle feeders, you can also turn feeding time into a game by tossing individual biscuits from your cat's dry diet across the room for him to chase and devour. If your cat has outdoor access, scattering his dry food across the lawn or yard can be a brilliant way of letting him search for his food. After he has finished, it's a good idea to check that he has found all the food, to prevent unwanted visitors to your garden. Also you should keep a check on how much food your cat is getting in this way, and reduce his daily rations accordingly.

You may wish to start the scattering with just a few biscuits, making sure that you place them close to your cat and they are easily visible. Once your cat becomes adept at this way of feeding, you can scatter his entire ration over a wider area—you will find that soon your cat will rarely miss a single biscuit. Scatter feeding is a good option on tiled or wooden floors, as the biscuits will bounce in all directions when scattered. However, if you have more than one cat, you will need to separate them to scatter feed to prevent one cat pinching the other cat's biscuits, and also so that the cats do not feel they have to compete over food.

Another nice way to provide a hunting game during feeding is to feed the cat's biscuits one by one into a long cardboard tube held at an angle, against a tiled or wooden floor, allowing the biscuits to roll out one end. Your cat will soon become adept at catching the rolling biscuit under one paw. You have in effect created your cat's very own game of Splat-A-Rat—in this case the "rat" is the biscuit and the cat needs no mallet as he is expertly skilled at quickly swiping the biscuit as it rolls past. This is even a game you can play sitting on the floor watching television or while on the phone—while you relax, your cat can have plenty of fun catching his supper.

Much of the time cats spend hunting is actually spent investigating the environment in the hope of locating signs of prey. Sensory boxes can provide an opportunity to explore without the need to leave home. These are simply large cardboard boxes that contain a host of items your cat might experience outside, some of which have

Cosmos and Sarah play Splat-A-Rat.

traces of the scent of prey. These can be items you have collected
from your garden or from walks, as well as household items. Dried
leaves, feathers, pebbles, grass, bark chippings, branches, twigs and
herbs can all be placed in cardboard boxes, some large enough for
your cat to jump into and have a good rummage around. Plastic
balls, as found in children's ball pits, can add interest: I have seen
cats literally "having a ball" as they toss these out of a box.

Adding to the box some of your cat's favorite toys, such as a fur
mouse or feathered toy, allows your cat to really "hunt" for his
"prey." My cats have all had some toy mice that emit a cheeping
sound when touched. They go wild for these in sensory boxes as
they can hear the cheep when they accidentally move the toy but
cannot immediately locate it. It gives them many minutes of fun as
they locate their toy. Herbie would often jump out of his larger sensory box with his catch in his mouth. For the less confident cat,
small sensory boxes can be made for rooting around with a single
paw: they can be created out of empty tissue boxes or other small
cardboard boxes. Although plastic boxes can be used, cardboard is

excellent at absorbing scent and has the added bonus of being able to be chewed. Thus, the box itself becomes part of the sensory experience. Tossing some food treats or your cat's ration of biscuits into the sensory box can be a great way to encourage your cat to "hunt" for his food as the treats or biscuits disperse and need to be individually found—in small tissue-sized boxes, the biscuits often have to be extracted from the box with a paw, in the same way as in some puzzle feeders.

Some cats who have a keen desire to hunt but not much opportunity (either through "hunting" games or access to real prey) will direct their predatory behavior toward people in the home. Hands and feet are common targets because they tend to be waved about and move rapidly during such everyday activities as conversations and walking. Some people subconsciously swing or tap a foot while resting, and such motion is often too irresistible for a cat to ignore. "Predatory" attacks on human flesh can indeed hurt, and often the target's response is an involuntary high-pitched squeal followed by much jumping out of the way of the cat. However, rather than

Herbie explores a homemade sensory box.

deterring the cat, such a response may actually reinforce the behavior, with the cat interpreting the faster movement and the high-pitched sounds as characteristic of prey.

"Predation" on our hands and feet is unlikely to occur if the cat has ample opportunity for appropriate hunting games. However, for cats who have already developed the habit, try increasing the frequency and diversity of hunting games and redirecting the cat on to more appropriate targets such as wand toys whenever he shows the first signs of predatory-like behavior directed toward you (wide stare, fixed eyes, crouched stalking position). Reducing the reinforcement properties of the target can also help—this may mean that for a while you need to wear gloves or Wellington boots or even chaps in the house, so you can keep still and are not tempted to squeal or flinch when an offending swipe or bite comes your way. Soon your cat will learn that toys are much more exciting "prey" than are hands or feet.

For cats with outdoor access, maintaining some control over your cat's behavior while outdoors can help prevent a successful hunt. Many of the training tasks in Chapter 10—training your cat to wear a harness for outdoor walks, encouraging him to walk with you outdoors and even using your recall training will help your cat to still be able to enjoy time outside but with reduced risk to the resident wildlife.

Some owners are now resorting to devices to help prevent successful hunting. The most common is a bell: as the cat moves, the bell attached to the collar jingles, with many believing such a sound alerts the prey to the proximity of the cat, giving it time to escape. Most cats seem untroubled by such an addition; however, the more sensitive cat may find its sound or the sensation of it dangling on his chest worrisome—such cats need to be trained to fully accept it.

As with any new object, start by placing the bell on the floor, and reward your cat for any exploration of it. Once your cat is fully comfortable with it being on the floor, you can pick it up and let it jingle quietly in your hand. Again, reward calm or inquisitive behavior. Then over several repetitions, jingle the bell and provide a reward—we want to teach the cat not to be perturbed by the sound. If at any stage your cat appears tense, you will need to go back to the earlier stages of training.

Once your cat is neither interested in nor bothered by its sound, we can proceed to attaching the bell to his collar—this is best done with the collar off the cat, as it can be a bit fiddly. Once the collar, with bell, is reattached to your cat, provide plenty of reward, either in the form of a meal or a game. Your cat will feel the bell and hear it every time he moves, but if he is rewarded with play or food intermittently, he will quickly learn that the bell is nothing to be worried about. At this stage, rewards are no longer needed as we want the cat to habituate to the sound of the bell.

Bells don't always work, so alternatives are coming onto the market. Many owners report that their cats still hunt successfully even with a bell attached to their collar—some have even reported that they have witnessed their cat carrying the bell in his mouth while hunting, presumably because the cat has found that this reduces the noise the bell makes. In light of this, there are now a handful of devices that attach to a standard cat collar, specifically designed to prevent or seriously reduce successful hunting attempts. One such device is a neoprene bib that attaches with Velcro to the cat's regular collar and hangs down his chest, obscuring the upper part of the front legs. It is not fully understood whether the claimed reduction in hunting success is due to the bright color of the bibs giving prey earlier warning of the cat's presence or whether the effectiveness of the cat's pounce is reduced because the bib gets in the way. Early research has suggested that bibs are better at preventing kills when the prey are birds, as opposed to small mammals, suggesting that it may indeed be the color (birds have even better color vision than we do). A note of caution, however: it is not fully understood how a dangling bib affects each individual cat's mobility, and thus for cats living close to roads, careful consideration should be given before trying this option.[7]

Another device is a brightly colored sleeve of fabric that encases the cat's collar. Again, research suggests that it reduces the number of birds killed by cats, the proposed mechanism being that the bright colors make the cat more visible to the bird, giving it more time to escape.[8] There are also more complex devices, still in prototype at the time of writing, that fit onto the cat's collar and emit a brief pulse of noise and short flash of light when the cat jumps, combining an audio and visual warning to the prey that a cat is nearby.

For the sake of the cat's well-being, all these devices require that the cat be trained to accept them. Training to wear the fabric collar sleeve and the light- and sound-emitting device should follow the same path as training your cat to wear a collar and a tracking device (Chapter 10), with the addition of acclimatizing your cat to the sounds and light flashes from the audio-visual deterrent—this can be done using the same training skills as used when acclimatizing a cat to the bell on its collar. The bib is very different from anything your cat will have encountered before because it dangles in front of his legs, and it may be more difficult to persuade your cat to accept it.

After being asked several times for advice on using bibs but having no experience of them, I began to watch videos of cats wearing the neoprene bibs. I was a little concerned about how much the bib might restrict cats' movement and therefore their safety and quality of life, and so, before forming an opinion and providing any advice to owners, I decided to try them with my cats: Cosmos and Herbie had never worn any hunting deterrents beyond a bell attached to their collars.

I began their training on a lovely sunny day when the cats were calmly lazing outside on the lawn. I started with Cosmos, as Herbie had ventured into the shade. As with anything new, I placed the bib a short distance from Cosmos to give him the opportunity to investigate it. He wasn't the least bit interested in it—neither worried nor intrigued by it—so I picked it up and offered it to him. He gave it a fleeting sniff and rolled over in the grass the way cats do when they are enjoying the sunshine. Because he was so relaxed, I decided it was a good time to try it on. Cosmos has always worn a breakaway collar and bell, and was trained to wear a number of tracking devices that attach to the collar in the past, so I knew the feeling of me attaching something to his collar while he was wearing it would not faze him. I dropped a treat next to Cosmos to encourage him to stand up, which he promptly did. I quickly but gently passed the piece of neoprene under his collar and secured it by pressing the Velcro together. I followed this by dropping a few more treats on the lawn—which due to the length of the grass took Cosmos a few seconds to find. This impromptu foraging game gave him time to feel the weight of the bib (which was actually very light) and the touch of it against his fur without being overly distracted by it—his mind was very much on finding his treats.

I next lured Cosmos for a few steps across the lawn using food to attract him. Cosmos adores training games and his mind was so focused on the lure, he did not seem worried about the bib, although his steps were much higher and more pronounced, as if he was trying to walk over the bib—bizarrely, it reminded me of the gait of a dressage pony! However, it did not take long at all—about a couple of minutes—before Cosmos realized he could walk normally and the bib would move out of the way of his legs. After rewarding Cosmos for following the lure several times, I let him play with a wand toy to give him the confidence to move a little faster with the bib. I interspersed short periods of play with periods of rest, thereby making sure the toy was not always a distraction to the bib but, instead, that Cosmos learned that wearing the bib led to rewards such as play while showing him he could still move freely while wearing the bib. Cosmos chased, swiped and jumped at the toy—all while wearing the bib. In fact, several times he caught the feather toy on the end of the wand—perhaps he will be one of those cats who outsmarts the hunting deterrent! My plan was to remove the bib from Cosmos after the play but he sauntered back into the house, jumped up onto the sofa and promptly fell asleep, so I left it on him. Herbie's training went pretty much the same way, although he wasn't as adept at catching the wand toy. Although it is clear that with plenty of training, a cat can readily accept wearing an antihunting device such as a bib, I wouldn't let my cats wear the bibs unsupervised—for me, I would worry that the bib would get caught on something and the cat would end up in a tangle or that it might impinge on the finer aspects of his movement as he navigated through more complex environments. Thus, the effort required to train a cat to wear an antihunting device, for me at least, is better invested in redirecting the cat's predatory instincts.

GIVEN THE WORLDWIDE TREND FOR GREATER CONCERN FOR ANIMAL welfare, it is likely that animal welfare legislation will continue to take increasing account of the concept of providing our pets with opportunities to display their natural behavioral patterns. Although the environment we keep our cats in is, in many parts of the world, rapidly changing from what used to be their natural environment, with a solid understanding of cat behavior coupled with well-established training skills and an imaginative mind, we can do our very best to make sure that our cats are happy and healthy in our care.

Conclusion

You did it.
You did it?
No, you *both* did it.

TRAINING YOUR CAT CAN BE—INDEED SHOULD BE—A TRANSFORMING experience, for owner and for cat alike. Any well-established cat-owner relationship will have inevitably included a great deal of *learning*, but most of that will have gone on almost unnoticed and in the background, as each party has gradually shifted his or her behavior to suit the other. You will have come to know, perhaps without really being aware of it, what times of day your cat is most receptive to attention. Your cat will have learned how you behave when you feel like spending some time with him.

The "magic" that training brings is not because it involves mysterious processes: all the exercises we have described in this book rely on just the same kind of learning that your cat will have used when exploring your home and its surrounds. The "magic" comes from those processes becoming so much more transparent to you, through being more deliberate and thought through: the connection between reward and a change in behavior will have been revealed to you. You will have witnessed those joyous moments when your cat suddenly seems to "get" what the response is that you have been aiming for (though you may also have been through the frustration of trying again the next day, only to find he appears to have forgotten the lot!). You will have become more sensitive to your cat's body language, judging when is a good time to begin a training session and when not to, when it's time to persevere with a particular task and when it's time to stop and leave it for another day. Your cat, for his part, will have learned a lot more about you than he might ever have done from the typical take-it-or-leave-it kind of relationship that cats are supposed (mostly wrongly) to prefer. Training will have brought you together to an extent that could not have happened any other way. Moreover, because your cat will now see you as the source of much joy, he will have come to see you in a much more positive light than before.

A well-trained cat will also be a source of amazement to those— in the majority, of course—who have uncritically accepted the received wisdom that cats are untrainable. When TV presenter Liz Bonnin came to Sarah's house to meet Herbie and Cosmos, she lightheartedly asked, "Do you think I could get Cosmos to sit? I'm sure it won't work"—but it did, and at the first attempt! "And that, ladies and gentlemen," Liz announced to the camera in a tone that mixed admiration with genuine wonder, "is how you get your cat to sit." You will find, whether or not that was your intention, that you've become an ambassador for training cats, particularly if you've extended your training beyond the indoors. If you've trained your cat to come when called or to walk in a comfortable and relaxed manner on a harness and leash, you'll probably even attract the admiration—perhaps grudgingly—of some dog owners who may have less of a grasp of how best to train their pet than you have of yours. For those cats that do not have access to the outdoors, the results of your training to help them cope better at the veterinarian's is also likely to be met with marvel—as your cat remains calm and relaxed in the veterinary waiting room—by those with anxious, quivering pets. Here is a perfect opportunity to spread your new-found knowledge.

IN TRAINING YOUR CAT, YOU WILL ALSO HAVE IN A VERY REAL WAY become part of the future of cat ownership. The cat of tomorrow

will need to be very different from the cat of yesterday, and training.
is an essential component of that change, change that cannot be
avoided if cats are to continue to be both popular and accepted by
society as a whole, not just by those who love cats.

The way that we conceive of cats has already undergone a seismic
shift over the past half century. In the England of the 1950s, from
which come John's first memories of cats, most cats were regarded as
more or less interchangeable (apart from pedigree cats). Ordinary
moggies (mongrel cats) bred freely, producing far more kittens than
could ever find homes: drowning was sadly still a widely accepted
method of population control. If a pet cat went missing, its departure
was explained away as the inevitable consequence of the animal's
independent spirit, and another soon took its place, either a kitten
or an adult cat that "just turned up." Few put up "lost cat" posters
(unless the animal happened to be valuable in the monetary sense).

Nowadays, most people enter cat ownership with the expectation that the relationship will be lifelong. That relationship is undoubtedly much deeper, in the emotional sense, than the looser
bond that characterized cat ownership sixty years ago. Cats are now
almost expected to be like dogs: affectionate, loyal and faithful from
whelping box to pet cemetery. The problem is that most cats aren't
really like that. Under the skin (i.e., genetically) they have changed
little from the semi-independent mousers that their ancestors were
until the twentieth century. It's not the cat's fault. Evolution simply
doesn't work that way. Natural selection needs both time (hundreds
of generations) and a steady selection pressure of some kind. Fifty
generations (assuming that females breed when they're ten months
old—most can actually breed several months earlier than this) isn't
enough, and it's not clear what the selection pressure might be, because nonpedigree cats usually choose their own mates. Does a female cat know to sidle up to a tom and ask him, "Are your other
kittens good with people, sociable with cats and not inclined to
hunt endangered species?" Whatever the criteria unneutered females use to select males (and vice versa—and both are almost
entirely unknown), it's unlikely to be any of these.

Unusual among domesticated animals, most cats are the product of natural selection (albeit selection that's played out in environments largely constructed by people) and not deliberate breeding. Such breeding will undoubtedly have a part to play in creating the cat of the future, but unfortunately today's pedigree cats, delightful though many of them are, are unlikely to be the starting point.

Currently, cat breeds are judged primarily on their appearance, not their behavior, and there is no evidence to suggest that any are better suited to modern urban living than alley cats are. Indeed the whole history of cat breeding has been based on refining their looks, whereas it is only for the past 120 or so years that dog breeding, prior to that driven by function, has also concentrated on altering the look of the animals. For both animals, the focus on the show ring has been a genetic disaster. Cats have not suffered in this respect as much as dogs have because the breeding has been less extreme, but many generations of mating between close relatives has caused all manner of genetically based diseases to surface, and a few breeds (though fewer than in dogs) are actually characterized by debilitating mutations.[1]

Fortunately, there is still a great deal of genetic variability in today's cats: a considerable amount even within one location, for example, London or Los Angeles, substantially more if the cats of different continents are combined together. Cross-breeding between mongrel cats from one place and pedigree cats whose genes originated in another—for example, a Siamese from Manhattan with a Brooklyn alley cat—should generate a wide range of different cat personalities, far more diverse than would occur naturally in either of the populations from which the parents had come. There is no reason to suppose that any of the offspring would not make perfectly satisfying and happy pets. However, some would be better suited to modern lifestyles than others would be, and if bred from, these cats could form the basis for a new type of superpet, selected for its behavior and temperament rather than its appearance, in contrast to all current breeds (indeed, the individual cats might look quite different from one another).[2]

Biologists are now close to identifying the individual genes that affect cats' personalities the most. One recent study narrowed the search to just thirteen genes that affect the way that the brain and nervous system are constructed. Such breakthroughs raise the possibility that within a few years the genes that have influenced any individual cat's personality could be "read" from the minute amount of DNA sticking to the end of a few of his (or her) hairs: this would enable the ready selection of the most promising candidates for breeding. (Because much of a cat's personality comes from its experiences in kittenhood and adolescence, without such techniques it is currently difficult to determine what influence its genes have had, although these effects are undoubtedly powerful.)[3]

What should this new "breed" be like? Well, judging by the current difficulties that domestic cats are encountering, it should differ from the current "Mr. Average" cat in four distinct ways:

1. Better able to resolve conflicts with other cats, both those in its own home and those in the neighborhood
2. More accepting of human behavior and approaches from unfamiliar people (without going so far that the cat becomes vulnerable to abuse[4])
3. Disinclined to go hunting when well nourished
4. More resilient to changes in its surroundings—both the cats and people it interacts with, and the physical place where it lives

There is every reason to suppose that selective breeding will be able to bring out all four of these traits, because domestic cats already vary so much in how inclined they are to go hunting, how generally adaptable they are, and how sociable they are to both people and cats. In the long term, it may be possible and indeed desirable to add a fifth item to this wish list:

5. Easy to train

It is highly unlikely that changes in genetics will ever allow us to bypass the crucial role of early experience (socialization) in giving cats the skills they need to interact with people. Whatever the cat's genetics, it must never be forgotten that much of its personality is the result of the combined efforts of genes and experience, and that although the way these combine and interact may change in the future, both will continue to be important. Thus learning will always play a role, some of it (as today) happening spontaneously, the product of the cat's normal everyday experiences, but an increasing proportion resulting from more formal structures of training.

However, this is all for the future. In the short term, the only surefire way of altering a cat's instinctive reactions is through training. Until the cat's genome becomes more pet-like, this process will have to be repeated at every generation, because although mother cats do influence their kittens' behavior, the kittens' personalities have not matured when they leave their mothers at the typical eight weeks of age—and not even at the twelve or thirteen weeks at which many pedigree cats go to their new homes. However, because training should be fun for both parties, the need to start again from scratch (pun intended) with each and every kitten should not be a hardship but rather an enjoyable way of forging a close relationship with every cat that you bring into your life.

THE PAST HALF CENTURY HAS SEEN THE BIRTH AND EBULLIENT growth of feline medicine (previously, believe it or not, the veterinary profession treated cats as small dogs). As a consequence, the cat's medical needs are now well catered to. Their diseases are understood as never before. A wide range of feline vaccines has appeared, protecting our pets from diseases that previously many cats would have succumbed to. Neutering has freed female cats from the burden of giving birth and raising kittens year after year after year, and male cats from having to compete with possibly more experienced or more agile rivals, anxious to keep them as far from "their" females as possible. From the owner's perspective, feline medicine has first empowered and then cemented the expectation that a cat is a pet for life, a permanent companion, and not a temporary liaison as heretofore.

There has been much less progress in veterinary and indeed scientific understanding of our pet cats' emotional and psychological needs. We now have the wherewithal to keep them physically healthy, but their subjective experiences are still something of a mystery to owner and professional alike. Our hope is that through this book, we have been able to demonstrate the benefits that training can provide to cats by making them better able to cope with the stressors that inevitably emerge from everyday life. However, these are not the only ways in which training can make a cat's life a happier one. For cats whose exposure to stress has already given rise to a behavioral disorder, training can play an important role in their resolution (see the Further Reading section for some suggested texts).

Training can also help to channel behavior that owners find undesirable—behavior that is natural insofar as the cat is concerned but annoys the owner (so not a behavioral disorder as such). Perhaps the behaviors that attract the most controversy are killing wildlife and scratching furniture. As we've seen in this book, there are ways to train a cat to direct such behaviors onto more acceptable targets.

WHETHER WE CHOOSE TO TRAIN THEM OR NOT, CATS WILL ALWAYS have inquisitive minds and so will go on learning every day of their lives, even though their behavior is at its most flexible during kittenhood and gradually becomes more habitual with age. You should think of training not as interfering with your cat's freedom but as a way of channeling his desire to learn toward making his life, and your relationship with him, as good as it can be. We owe this to our cats: they have come a long way in turning themselves into an animal that likes to please us; training is the best way we have to help them complete that journey.

Acknowledgments

John: First and foremost, I thank Sarah for coming up with the initial idea for this book and for sharing all her expertise, not just on the nuts and bolts of how to train cats, but also her understanding of how learning theory can be applied in molding the behavior of an animal with an undeserved reputation for inflexibility. Also, much gratitude to Dr. Debbie Wells of the Queens University Belfast who invited me to examine Sarah's PhD thesis and thereby introduced us to one another— and to Helen Sage for asking us both to join the BBC Horizon team in Shamley Green making *The Secret Life of the Cat*, where we hatched the idea for this book between filming sessions.

I wish to thank all the former and current academic colleagues who have assisted me in my research and in the writing of my earlier book *Cat Sense*, where they are listed by name. Also my family for allowing me the time to complete another book about cats.

Sarah: I owe much thanks to John for believing in my early work on training cats, for encouraging me to take the leap to set down these ideas more formally and most importantly, for joining me on this writing journey— presenting a book that is both a practical guide and a factual account of the cat has had its challenges. It has been very much the case that "two brains are better than one"—writing alongside John has made it a very rewarding process. I am so grateful for all the scientific discussions John and I have had over the years—they have helped me greatly develop my knowledge of the cat, and I have enjoyed being able to share ideas and discuss new concepts.

Many scientists, clinical animal behaviorists and dog trainers have helped me develop my ideas about training cats through letting this "cat person" into their lectures, training classes and discussions—you have really helped me think laterally. They include Professor Daniel Mills, Helen Zulch, Dr. Debbie Wells, Dr. Oliver Burman, Dr. Hannah Wright, Chirag Patel, Jessica Hardiman, Jo-Rosie Haffenden and Hannah Thompson. I have been incredibly fortunate to have had numerous opportunities over the years to discuss feline behavior with many specialists including Vicky Halls, Sarah Heath, Nicky Trevorrow, Dr. Lauren Finka, Rachael King, Naima Kasbaoui, Claire Bessant, Dr. Andy Sparkes and Dr. Rachel Casey—these discussions have greatly shaped my current views. I am also very grateful to those on the other side of the pond who have helped me learn what it is like to be an American cat—they include Ilona Rodan, Mikel Delgado, Julie Hecht, Theresa DePorter, Miranda Workman, Jacqueline Munera and Steve Dale. Likewise, a huge thanks to all working at International Cat Care—the charity was the first organization to ask me to share my ideas on training cats with an invitation to talk on the subject at its annual conference in 2010 and has continued to support the concept.

Heartfelt thanks go to my cats, who have been constant companions during the extended periods I have been behind my computer, and to my husband, Stuart, and my parents, who have all supported me unconditionally throughout the writing of this book, both in time and love.

John and I are very grateful to Peter Baumber for being such a wonderful cat photographer—his understanding of feline behavior coupled with his excellent photography skills meant our model cats were not only always relaxed in front of the camera but were even comfortable enough to demonstrate their training abilities. Thanks goes to Emma Schmitt (owner of Batman), Lizzie Malachowski (owner of Sheldon) and to the charity Lincs Ark (who were caring for Skippy at the time) for kindly allowing us to photograph some of their cat's training skills.

We both wish to thank our agent Patrick Walsh of Conville & Walsh and his staff, and our editors at Basic Books, Lara Heimert and Roger Labrie, for helping us to mold our ideas into a form which we sincerely hope that cat owners will find both accessible and enjoyable.

Further Reading

John's *Cat Sense* (New York: Basic Books, 2013) provides an accessible account of domestication and cat behavior that will complement the introductions to the various chapters in this book. More detailed accounts, suitable for serious students of cat behavior, are available in the second edition of John's textbook, coauthored by Drs. Sarah Brown and Rachel Casey, *The Behaviour of the Domestic Cat* (Wallingford, UK: CAB International, 2012), and also in the three editions of *The Domestic Cat: The Biology of Its Behaviour*, edited by Professors Dennis Turner and Patrick Bateson and published by Cambridge University Press in 1988, 2000 and 2013.

Those readers whose interest in cats ranges wider than the one in their lap might try Andrew Kitchener's *The Natural History of the Wild Cats* (Ithaca: Cornell University Press, 1997) or Fiona and Mel Sunquist's *The Wild Cat Book: Everything You Ever Wanted to Know About Cats* (Chicago: University of Chicago Press, 2014).

For a general account of animal training, *Carrots and Sticks: Principles of Animal Training* (Cambridge: Cambridge University Press, 2008) by Professors Paul McGreevy and Bob Boakes is hard to beat, but also increasingly difficult to find. John M. Pearce's *Animal Learning and Cognition: An Introduction*, 3rd edition (New York: Routledge, 2008), is one of the best undergraduate-level texts on the same topic.

For those introducing a dog to a cat (or cats) or vice versa, Sarah Fisher and Marie Miller's *100 Ways to Train the Perfect Dog* (Newton Abbot, UK: David & Charles, 2010) describes how to train dogs in many of the key skills needed to help ensure smooth introductions with cats.

If you are seeking advice for a behavior issue with your own cat that's not covered in this book, there is no substitute for a one-on-one consultation with a qualified expert, although these are not yet available in all areas (most "behavior consultants" specialize in dogs). In some countries, there are professional organizations for qualified cat behavior experts (some of whom are also veterinarians)—for example, in the United States there is the American Veterinary Society of Animal Behavior (www.avsabonline.org) and the International Association of Animal Behavior Consultants (www.iaabc.org). Books by Sarah Heath and Vicky Halls, including *Cat Detective: Solving the Mystery of Your Cat's Behaviour* (London: Bantam Books, 2006), and Pam Johnson-Bennett may provide some useful advice. In addition, the International Society of Feline Medicine (ISFM) has produced the *ISFM Guide to Feline Stress and Health: Managing Negative Emotions to Improve Feline Health and Wellbeing*, ed. Sarah Ellis and Andy Sparkes (Tisbury, UK: International Cat Care, 2016), which describes the different negative emotions cats can experience, what can cause them and how to prevent or alleviate them.

Linda Case's book *The Cat: Its Behavior, Nutrition and Health* (Ames: Iowa State Press, 2003) is a good introduction to the domestic cat's needs, and includes a chapter on how cats learn as well as one on problem behaviors. Feline welfare is covered briefly in the third edition of *The Domestic Cat: The Biology of Its Behaviour*, but more comprehensively in *The Welfare of Cats*, edited by Irene Rochlitz (Dordrecht, the Netherlands: Springer, 2007).

Notes

Notes to Preface

1. See the American Veterinary Medical Association's position on this issue at https://www.avma.org/Advocacy/StateAndLocal/Pages/ownership-vs-guardianship.aspx.

Notes to Introduction

1. The evolution of the cat family has recently been revised based on differences among the species in their DNA, revealing some remarkable migrations. See Stephen O'Brien and Warren Johnson's article "The Evolution of Cats" in *Scientific American* (July 2007), 68–75.

2. More detail on the domestic cat's recent ancestors, and domestication itself, can be found in Chapters 1 and 2 of *Cat Sense* (see Further Reading). See also the article by Carlos Driscoll, Juliet Clutton-Brock, Andrew Kitchener and Stephen O'Brien, "The Taming of the Cat," in the June 2009 issue of *Scientific American*, pages 56–63.

3. Chapter 7 of *Cat Sense*, "Cats Together" (see Further Reading), has a more comprehensive account of our understanding of cats' social lives; see also Chapters 5 and 6 of the third edition of *The Domestic Cat: The Biology of Its Behaviour*.

4. The transformation of the domestic cat from hunter to companion and then to object of worship is described in Jaromir Malek's book *The Cat in Ancient Egypt* (London: British Museum Press, 1996).

5. The charity International Cat Care maintains an up-to-date database of inherited conditions among breeds in the UK at http://icatcare.org/advice/cat-breeds/inherited-disorders-cats.

6. The unique body plan of the cat family, and how it evolved, is described in Andrew Kitchener's book *The Natural History of the Wild Cats* (see Further Reading).

7. As with human nutrition, the feeding of cats has been subject to a considerable degree of "faddism" in recent years. The scientific viewpoint is described in a free booklet, *The WALTHAM Pocket Book of Essential Nutrition for Cats & Dogs*, downloadable from http://www.waltham.com /resources/waltham-booklets/.

8. The experiments that established this link between play and hunting, conducted by John and his colleague Dr. Sarah Hall, are described in Chapter 6 of *Cat Sense* (see Further Reading).

9. The Smithsonian Conservation Biology Institute has published information comparing predation by house cats and feral cats: http://www .nature.com/ncomms/journal/v4/n1/full/ncomms2380.html.

10. Cats' senses are described in more detail in John's books, *The Behaviour of the Domestic Cat* (Chapter 2) and *Cat Sense* (Chapter 5).

11. For a recent discussion of the evidence for how the human brain evolved, see Robin Dunbar's *Human Evolution* (London: Pelican Books, 2014).

12. Some scientists believe that domestic dogs may have a rudimentary theory of mind, but dogs are much more attentive to us than cats are. For a discussion of the possibilities, see Alexandra Horowitz's paper "Theory of Mind in Dogs? Examining Method and Concept" in the journal *Learning & Behavior* 39 (2011): 314–317.

13. Neuroscientist Gregory Berns has written an engaging account of how he trained his dog Callie to sit in an MRI scanner: *How Dogs Love Us* (Seattle: Lake Union, 2013).

14. There are several accessible in-depth accounts of how dogs' minds work, including Alexandra Horowitz's *Inside of a Dog: What Dogs See, Smell and Know* (New York: Simon & Schuster, 2009); John Bradshaw's *Dog Sense* (New York: Basic Books, 2011) and *In Defence of Dogs* (London: Penguin, 2012); and Brian Hare and Vanessa Woods's *The Genius of Dogs* (New York: Plume, 2013).

15. For studies of the effects of viewing pictures of kittens, see papers by Hiroshi Nittono, Michiko Fukushima, Akihiro Yano and Hiroki Moriya, "The Power of Kawaii: Viewing Cute Images Promotes a Careful Behavior and Narrows Attentional Focus," *PLoS ONE* 7, no. 9 (2012): e46362, doi:10.1371/journal.pone.0046362; Andrea Caria and colleagues, "Species-Specific Response to Human Infant Faces in the Premotor Cortex," *NeuroImage* 60 (2012): 884–893.

16. Stress-induced behavioral disorders in cats are described by veterinarian Dr. Rachel Casey in Chapters 11 and 12 of *The Behaviour of the Domestic Cat* (see Further Reading).

Notes to Chapter 1

1. The science of animal learning can make for a daunting read: two of the more accessible texts are McGreevy and Boakes's and John M. Pearce's books (see Further Reading).

2. Textbooks that deal with the theory of animal learning have their own terminology for these processes, such as "positive reinforcement" and "negative punishment" (the latter corresponding to Consequence 2), but most people find these confusing and so we have not used them here.

3. For more detail of this experiment, see Chapter 6 of John's *Cat Sense*.

4. For more information on how and when cats like to feed, see John and Chris Thorne's chapter "Feeding Behaviour" in *The Waltham Book of Dog and Cat Behaviour*, ed. Chris Thorne (Oxford: Pergamon Press, 1992), 115–129.

5. In an unpublished study by Sarah and her colleagues, cats played with toys on wands (that differed in their sensory properties) to investigate the effects such properties had on encouraging cats to play. What became quite apparent was that cats were much more likely to chase the toy, regardless of which toy it was, if the wand was moved in quick, straight lines as opposed to moved slowly, randomly, or in a to-and-fro movement.

6. This blog post provides a write-up of the study Sarah conducted that examined cats' behavioral responses to being touched by both owners and strangers on different parts of the body: http://www.companionanimalpsychology.com/2015/03/where-do-cats-like-to-be-stroked.html. The social significance of the various signals is discussed in more detail in Chapters 7 and 8 of John's *Cat Sense*, and the chapter by Sarah Brown and John Bradshaw in the third edition of Turner and Bateson's *The Domestic Cat* (see Further Reading).

Notes to Chapter 2

1. Behavioral differences among cat breeds are discussed further by Benjamin Hart, Lynette Hart and Leslie Lyons in their chapter in the third edition of Turner and Bateson's *The Domestic Cat* (see Further Reading).

2. The origins of feline personalities are discussed further in Chapter 9 of John Bradshaw's *Cat Sense*, and by Michael Mendl and Robert Harcourt in the second edition of Turner and Bateson's *The Domestic Cat* (see Further Reading).

3. For some background on how previous experiences, especially negative ones, may affect a cat's readiness for training, see chapters by Irene Rochlitz, entitled "Feline Welfare Issues," and by Judi Stella and Tony Buffington, "Individual and Environmental Effects on Health and Welfare," in the third edition of Turner and Bateson's *The Domestic Cat* (see Further Reading).

Notes to Chapter 3

1. For an account of how the controllability of a situation is thought to influence an animal's welfare, see Daniel Mills, ed., *The Encyclopedia of Applied Animal Behaviour and Welfare* (Wallingford: CAB International, 2010), 136–137.

2. For further information on desensitization and counterconditioning, see Pamela Reid's book *Excel-erated Learning: Explaining in Plain English How Dogs Learn and How Best to Teach Them* (Hertfordshire: James & Kenneth, 1996), 150–152. While the title suggests this book is just about dogs, it does provide a very good overview of learning theory that is just as applicable to cats.

3. Harry was Sarah's cat, who before coming to live with her had lived in an outdoor pen.

4. See www.clickertraining.com.

5. The way that links form between emotion and reinforcers is explained in Edmund Rolls's article "Neural Basis of Emotions" in the *International Encyclopedia of the Social & Behavioral Sciences*, 2nd ed., vol. 7, edited by James D. Wright (Oxford: Elsevier, 2015), 477–482.

6. For more on how cats use their sense of smell, and their vomeronasal organ, to acquire information about their surroundings, see Chapter 5 of John's *Cat Sense* (see Further Reading).

7. More information about B. F. Skinner, schedules of reinforcement and details of his book can be found at the B. F. Skinner Foundation website www.bfskinner.org.

8. See Bonnie Beaver's book *Feline Behavior: A Guide for Veterinarians*, 2nd ed. (St Louis, MO: Saunders, 2003), 68–69.

Notes to Chapter 4

1. Several studies were carried out in Eileen Karsh's laboratory in the 1980s, where kittens received different types and amounts of handling at different ages. From these data, the period most sensitive to handling was determined, as well as the type of handling that would have the most positive influences for later life. A summary of these studies can be found in Dennis Turner's chapter entitled "The Human-Cat Relationship" in the third edition of Turner and Bateson's book *The Domestic Cat: The Biology of Its Behaviour* (see Further Reading).

2. This study is published as "Human Classification of Context-Related Vocalizations Emitted by Familiar and Unfamiliar Domestic Cats: An Exploratory Study," by Sarah Ellis, Victoria Swindell and Oliver Burman in the journal *Anthrozoös* 28 (2015): 625–634.

3. See note 1 above.

4. The social behavior of cats is discussed further in chapter 8 of *Cat Sense* (see Further Reading).

5. The areas on the body where cats respond most positively to touch were recently investigated by Sarah and colleagues and published as "The Influence of Body Region, Handler Familiarity and Order of Region Handled on the Domestic Cat's Response to Being Stroked," by Sarah Ellis, Hannah Thompson, Cristina Guijarro and Helen Zulch in *Applied Animal Behaviour Science* 173 (2016): 60–67. This blog post provides a write-up of the study: http://www.companionanimalpsychology.com/2015/03/where-do-cats-like-to-be-stroked.html.

Notes to Chapter 5

1. The social structure of free-ranging cats is discussed in greater detail in Chapter 8 of John's book *Cat Sense* (see Further Reading).

2. See Chapter 4, note 1.

3. As part of her doctoral work, Sandra McCune demonstrated that the friendliness of the father had a positive impact on the kitten's ability to cope with novel objects and novel people. It is highly likely that this effect also extends to a kitten's view of other cats. For a summary of her work, see Michael Mendl and Robert Harcourt's chapter entitled "Individuality in the Domestic Cat: Origins, Development and Stability," in the third edition of Turner and Bateson's book *The Domestic Cat: The Biology of Its Behaviour* (see Further Reading).

4. Comprehensive guidelines (written by Sarah and an international team of feline experts) on how to set up a cat's resources within the home to optimize its well-being have been jointly produced by the American Association of Feline Practitioners and the International Society of Feline Medicine and can be freely accessed using the following link: http://jfm .sagepub.com/content/15/3/219.full.pdf+html.

5. More information about the cat's sensory world can be found in Chapter 5 of John's book *Cat Sense* (see Further Reading).

Notes to Chapter 6

1. The Pet Food Institute provides the most recent statistics on dog and cat pet populations in the United States of America. They can be found at http://www.petfoodinstitute.org/?page=PetPopulation.

2. Naturalist Mike Tomkies provides a graphic description of just how untameable wildcats can be in *My Wilderness Wildcats* (London: Macdonald & Jane's, London, 1977).

3. The age of introduction and whether to introduce a dog to the resident cat, or vice versa, was the subject of a research study: N. Feuerstein and J. Terkel, "Interrelationships of Dogs (*Canis familiaris*) and Cats (*Felis catus L.*) Living Under the Same Roof," *Applied Animal Behaviour Science* 113, no. 1 (2008): 150–165.

4. For more information on how to teach the dog these skills, see Sarah Fisher and Marie Miller's *10 Ways to Train the Perfect Dog* (Newton Abbot, UK: David & Charles, 2010).

Notes to Chapter 7

1. See Irene Rochlitz's chapter "Feline Welfare Issues" in the third edition of Dennis Turner and Patrick Bateson's *The Domestic Cat: The Biology of Its Behaviour*, 131–153 (see Further Reading).

2. An overview of various animals' ability to smell fear can be found in the introduction of a research study investigating whether humans can detect fear in other humans from scent alone conducted by Kerstin Ackerl, Michaela Atzmueller and Karl Grammer and published as "The Scent of Fear" in *Neuroendocrinology Letters*. A free-access online version of the article can be found at http://evolution.anthro.univie.ac.at/institutes /urbanethology/resources/articles/articles/publications/NEL230202R03 scent.pdf.

Notes to Chapter 8

1. The late Penny Bernstein wrote a chapter entitled "The Human-Cat Relationship" in Irene Rochlitz's book *The Welfare of Cats* (see Further Reading), which documents many of the benefits cats bring to humans. Stroking is usually enjoyed by cats, but a study conducted by Daniel Mills and his colleagues showed that for cats who were not fully relaxed in their home, stroking could provide an additional stressor. A summary of this study can be found on the news pages of University of Lincoln's webpage: http://www.lincoln.ac.uk/news/2013/10/772.asp.

2. For a detailed description of kitten development, see Chapter 4, "Every Cat Has to Learn to Be Domestic," of John's book *Cat Sense* (see Further Reading).

3. A greater description of colony scents, including how the badger creates a colony scent, can be found in Chapter 5 entitled "Communication" in the second edition of John's book *The Behaviour of the Domestic Cat* (see Further Reading).

4. The areas on the body where cats respond most positively to touch were recently investigated by Sarah and colleagues and published as "The Influence of Body Region, Handler Familiarity and Order of Region Handled on the Domestic Cat's Response to Being Stroked," by Sarah Ellis, Hannah Thompson, Cristina Guijarro and Helen Zulch in *Applied Animal Behaviour Science* 173 (2015): 60–67, http://dx.doi.org/10.1016/j.applanim.2014.11.002. The following blog post provides a write-up of the study Sarah conducted that examined cats' behavioral responses to being touched by both owners and strangers on different parts of the body: http://www.companionanimalpsychology.com/2015/03/where-do-cats-like-to-be-stroked.html.

Notes to Chapter 9

1. Cats have a range of behavioral strategies for dealing with stress. These are described in great detail in the International Society of Feline Medicine *Guide to Feline Stress and Health: Managing Negative Emotions to Improve Feline Health and Wellbeing* (see Further Reading).

2. The *International Society of Feline Medicine Guide to Feline Stress and Health: Managing Negative Emotions to Improve Feline Health and Wellbeing* provides detailed information on the impact of stress on a cat's physiology (see Further Reading).

3. A description of the behavioral signs of frustration, based on a study in a Canadian shelter, can be found in N. Gourkow, A. LaVoy, G. A. Dean and C. J. Phillips, "Associations of Behaviour with Secretory Immuno-globulin A and Cortisol in Domestic Cats During Their First Week in an Animal Shelter," *Applied Animal Behaviour Science* 150 (2014): 55–64. Information on the effects chronic stress can have on a cat's health can be found in the *ISFM Guide to Feline Stress and Health: Managing Negative Emotions to Improve Feline Health and Wellbeing* (see Further Reading).

4. The role of cortisol in stress is covered in detail in the *ISFM Guide to Feline Stress and Health: Managing Negative Emotions to Improve Feline Health and Wellbeing* (see Further Reading).

5. Sarah Heath and Vicky Halls's book *Cat Detective* may provide some clues as to what may be wrong (see Further Reading), as might the website of International Cat Care (www.icatcare.org), but if in any doubt, you should seek advice from your veterinarian, who will be able to refer you to a qualified behaviorist if needed. Benjamin and Lynette Hart's chapter "Feline Behaviour Problems and Solutions," in the third edition of Dennis Turner and Patrick Bateson's book *The Domestic Cat: The Biology of Its Behaviour* provides a succinct overview (see Further Reading). Detailed descriptions of the different negative emotional states can be found in Chapter 2 of Daniel Mills, Maya Braem Dube and Helen Zulch's book entitled *Stress and Pheromonatherapy in Small Animal Clinical Behaviour* (Chichester, UK: Wiley-Blackwell, 2012), 37–68.

6. Cat Friendly Clinic is a worldwide program from the International Society of Feline Medicine, the veterinary division of International Cat Care (and is known as Cat Friendly Practice in the United States and is run by the American Association of Feline Practitioners). It is designed to help veterinary practices make their clinics more cat friendly, thereby reducing the stress for the cat and making veterinary visits easier for cat owners as well. More information can be found at www.catfriendly clinic.com.

7. The American Association of Feline Practitioners and the International Society for Feline Medicine (part of the charity International Cat Care) have jointly produced best-practice guidelines entitled *Feline Friendly Handling*, which are published in the *Journal of Feline Medicine and Surgery*. These will not only help you with handling training at home in preparation for a vet visit but also give you a standard to which your own veterinary practice staff are adhering. They are freely available from guide-lines.jfms.com.

Notes to Chapter 10

1. The American Association of Feline Practitioners has a position statement supporting keeping pet cats as indoor-only cats. It can be found at http://www.catvets.com/guidelines/position-statements/confinement -indoor-cats.

2. Chapter 10 of John's book *Cat Sense* (see Further Reading) discusses the impact of cats on wildlife. In the BBC Horizon TV documentary entitled *The Secret Life of the Cat*, the hunting behavior of fifty UK cats (with ample opportunity to hunt in a wildlife-rich area) was monitored over the course of a week by John and Sarah. Very few prey were collected in the week by the cats' owners—in fact, numbers equated to less than two prey items per cat, suggesting successful kills were actually low. More details of the documentary can be found at http://www.bbc.co.uk/programmes/b02 xcvhw.

3. For a fuller description of how free-ranging cats maintain their territories, see Sarah Brown and John Bradshaw's chapter entitled "Communication in the Domestic Cat: Within—and Between—Species," in the third edition of *The Domestic Cat: The Biology of Its Behaviour* (see Further Reading).

4. The Australian government's Department of Education's website holds information pertaining to prey that (feral) cats most commonly hunt in that country: https://www.environment.gov.au/biodiversity/invasive -species/feral-animals-australia/feral-cats.

5. Guidelines detailing a cat's environmental needs, including their need for environmental complexity and opportunity for cat-specific behaviors such as exploration and hunting, have been produced by the International Society of Feline Medicine and the American Association of Feline Practitioners: http://jfm.sagepub.com/content/15/3/219.full.pdf+html.

6. A study carried out by Kathy Carlstead and colleagues on laboratory cats showed that daily unpredictable events that they could not control caused them to exhibit behavioral and physiological signs of stress. See K. Carlstead, J. L. Brown and W. Strawn, "Behavioral and Physiological Correlates of Stress in Laboratory Cats," *Applied Animal Behaviour Science* 38 (1993): 143–158.

7. The *ISFM Guide to Feline Stress and Health: Managing Negative Emotions to Improve Feline Health and Wellbeing* provides plentiful information on how to recognize, prevent and manage frustrated cats (see Further Reading).

8. Sacramento Leashwork's blog entitled "*Both Ends of the Leash*" focuses on dogs but has a very informative post on frustration bursts, of which the theory is entirely applicable to cats: https://leashworks.wordpress.com /2014/08/12/foundations-of-training-frustration-bursts/.

Notes to Chapter 11

1. The UK Animal Welfare Act (2006) stipulates under the duty of responsibility for a companion animal that the owner will ensure the animal has not only adequate physical welfare but also psychological welfare. Details of the act can be found at http://www.legislation.gov.uk/ukpga/2006 /45/contents. Information about US federal law regarding the keeping of pets, including cats, can be found on the website of the US Department of Agriculture's National Agriculture Library, which can be found at https:// awic.nal.usda.gov/government-and-professional-resources/federal-laws.

2. Sarah Brown and John Bradshaw's chapter entitled "Communication in the Domestic Cat: Within—and Between—Species," in the third edition of *The Domestic Cat: The Biology of Its Behaviour* (see Further Reading) provides a fuller overview of the ways cats use signaling to deter others from their territories. Urine-spraying can also occur for reasons of stress such as feeling threatened or anxious. If your cat is urine-spraying, the first point of call is to the veterinarian to check there are no medical problems. He or she may then decide to refer you to a qualified behaviorist for help. Further information on urine-spraying can be found in the "Guidelines for Diagnosing and Solving House-Soiling Behavior in Cats," produced by the American Association of Feline Practitioners and the International Society of Feline Medicine, published in the *Journal of Feline Medicine and Surgery*. They can be freely accessed at http://jfm.sagepub.com/content /16/7/579.full.pdf+html.

3. There is a more detailed discussion of the issues around surgical declawing in Chapter 11 of John's book *Cat Sense* (see Further Reading).

4. In John's book *Cat Sense* there is a whole chapter (10) dedicated to cats and wildlife (see Further Reading).

5. The catnip response is further described in John's book *Cat Sense* on pages 120–121 (see Further Reading).

6. For example, the UK Codes of Practice for pet owners that accompany the Animal Welfare Act (2006) stipulate that overfeeding of pets is a "serious welfare concern" that can lead to unnecessary suffering. Although breaking such codes does not equate to breaking the law, if an owner were

ever taken to court, failure to have adhered to the code of practice could
be used against him or her. The Code of Practice for the welfare of cats can
be found at https://www.gov.uk/government/publications/code-of-practice
-for-the-welfare-of-cats.

7. Information about the neoprene cat bib can be found at http://www
.catbib.com.au.

8. Details of the brightly colored bird-deterrent cuff can be found at
http://www.birdsbesafe.com.

Notes to Conclusion

1. The charity International Cat Care keeps a database of these disorders, applicable mainly to the UK: http://icatcare.org/advice/cat-breeds
/inherited-disorders-cats.

2. For a worldwide survey, see Monika Lipinski and colleagues' paper,
"The Ascent of Cat Breeds: Genetic Evaluations of Breeds and Worldwide
Random-Bred Populations," *Genomics* 91 (2008): 12–21.

3. Identification of some of the genes that affect the way the cat's ner-
vous system works and may have played a part in its domestication is de-
scribed in "Comparative Analysis of the Domestic Cat Genome Reveals
Genetic Signatures Underlying Feline Biology and Domestication," by
Michael Montague and coworkers, *Proceedings of the National Academy of
Sciences of the United States of America* 111 (2014): 17230–17235.

4. For example, there have been concerns that the most extreme forms
of the Ragdoll breed, which become limp and apparently insensible when
picked up, are at risk of being hurt either accidentally or even deliberately.
Such concerns were documented in the UK-based newspaper *Sunday Ex-
press* on December 11, 1994, of which an excerpt, reproduced by Sarah
Hartwell, can be found at http://messybeast.com/ultracat.htm.

Index

John Bradshaw is foundation director of the Anthrozoology Institute at the University of Bristol, and author of the *New York Times* best sellers *Cat Sense* and *Dog Sense*. He lives in Southampton, England. **Sarah Ellis** is feline behavior specialist at the charity International Cat Care and a visiting fellow at the University of Lincoln. She lives in Wiltshire, England.